Homesteading for Beginners

The Ultimate Self-Sufficiency Guide to Country Living, Raising Livestock and Natural Parasite Management

© Copyright 2024 - All rights reserved.

The content contained within this book may not be reproduced, duplicated, or transmitted without direct written permission from the author or the publisher.

Under no circumstances will any blame or legal responsibility be held against the publisher or author for any damages, reparation, or monetary loss due to the information contained within this book, either directly or indirectly.

Legal Notice:

This book is copyright-protected. It is only for personal use. You cannot amend, distribute, sell, use, quote, or paraphrase any part of the content within this book without the consent of the author or publisher.

Disclaimer Notice:

Please note the information contained within this document is for educational and entertainment purposes only. All effort has been executed to present accurate, up-to-date, reliable, and complete information. No warranties of any kind are declared or implied. Readers acknowledge that the author is not engaging in the rendering of legal, financial, medical, or professional advice. The content within this book has been derived from various sources. Please consult a licensed professional before attempting any techniques outlined in this book.

By reading this document, the reader agrees that under no circumstances is the author responsible for any losses, direct or indirect, that are incurred as a result of the use of the information contained within this document, including, but not limited to, errors, omissions, or inaccuracies.

Table of Contents

PART 1: COUNTRY LIVING 1
   INTRODUCTION 2
   SECTION ONE: SETTING UP YOUR HOMESTEAD 3
   CHAPTER 1: PLANNING YOUR HOMESTEAD 4
   CHAPTER 2: SUSTAINABLE LIVING CONSIDERATIONS 15
   CHAPTER 3: SHOULD YOU GO OFF-GRID? 22
   CHAPTER 4: BUILDING YOUR HOMESTEAD 25
   SECTION TWO: LIVING OFF THE LAND I: PLANT-BASED FOOD 32
   CHAPTER 5: GETTING STARTED WITH GARDENING EDIBLES 33
   CHAPTER 6: HARVESTING AND MAKING YOUR OWN PLANT-BASED FOOD 37
   CHAPTER 7: CANNING AND PRESERVING PLANT-BASED FOOD 52
   SECTION THREE: LIVING OFF THE LAND II: ANIMALS, MEAT, AND DAIRY 64
   CHAPTER 8: THE BUZZ ON BEEKEEPING 65
   CHAPTER 9: LIVESTOCK SELECTION AND CARE 80
   CHAPTER 10: MAKING YOUR OWN DAIRY AND MEAT-BASED FOOD 99
   CHAPTER 11: PRESERVING MEAT, DAIRY, AND EGGS 106
   BONUS CHAPTER: YOUR HOMESTEADING CHECKLIST 112
   CONCLUSION 117
PART 2: NATURAL PARASITE MANAGEMENT FOR LIVESTOCK 118
   INTRODUCTION 119

CHAPTER 1: WHAT ARE LIVESTOCK PARASITES? 121
CHAPTER 2: CLINICAL SIGNS AND DIAGNOSIS 136
CHAPTER 3: CHOOSING THE RIGHT NATURAL METHODS 152
CHAPTER 4: GRAZING THE PARASITES AWAY – PASTURE MANAGEMENT 161
CHAPTER 5: NUTRITION AND IMMUNITY 173
CHAPTER 6: HERBAL REMEDIES 187
CHAPTER 7: ADDITIONAL NATURAL STRATEGIES 206
CHAPTER 8: LIVESTOCK PARASITES AND CLIMATE CHANGE 217
GLOSSARY OF TERMS AND PARASITE REFERENCE 230
CONCLUSION 238
HERE'S ANOTHER BOOK BY DION ROSSER THAT YOU MIGHT LIKE 240
REFERENCES 241

Part 1: Country Living

The Ultimate Guide to Homesteading, Beekeeping, Raising Livestock, and Achieving Self-Sufficiency in the Countryside

Introduction

Country living . . . Homesteading. What images do these terms conjure up in your mind? "Little House on the Prairie" (for those old enough to remember it)? Low-slung farmhouses built of timber, surrounded by fields full of crops and livestock? You wouldn't be too far off the mark, but the reality of homesteading is a little less glamorous, although no less rewarding.

It's fair to say that homesteading isn't for the faint-hearted, nor is it for those who think it's an easy life and aren't prepared to put in the effort to make it work. It offers you a way to live your life free of the shackles of society, with little reliance on public utilities, and a way to produce your own food and many other things you rely on in your current life. It is hard work, but your effort will pay dividends in the long run.

If you aren't sure where to start or what to do, you've chosen the right book. It will lay out for you every step you need to take, everything you need to consider, and tips on making it all the success you dreamed it would be. This is written in plain English, with no hard-to-understand terms, plenty of images, and step-by-step instructions to help you along the way.

If you've dreamed of living a homestead life, settle in and read. By the end of this book, you will know everything you've ever wanted to know, and a bit more besides – and you'll know if this is the way of life for you or not.

Section One:
Setting Up Your Homestead

Chapter 1: Planning Your Homestead

Homesteads mean something different to everyone you speak to, but broadly speaking, it's about living self-sufficiently. Most people consider the most essential parts to be the land, the buildings on it, and farming to be self-sufficient, at least partly, if not totally.

So, what do you need to consider?

Location

Choosing your land and its location is the first place to start.
https://unsplash.com/photos/person-holding-red-round-medication-pill-Z8UgB80_46w

The obvious place to begin is the land, its location, and several other factors that you need to consider, and it all comes down to what you want.

Space and Size:

What size of land you want depends on what you want to do with it, but some things you need to consider are:

- **Your Family:** You need enough space for everyone joining you on your journey, so your land must be big enough for a big enough house.
- **Livestock:** How many animals are you planning to keep? You may want to start small, but remember, you need room to expand, too. Make sure you have enough land for housing and grazing.
- **Crops:** You need enough space to plant, harvest, and store your crops, as well as rotate your crop areas.

Also, think about the maintenance. The larger the land, the more maintenance it takes, so ensure you understand the time and resources needed.

What Are You Using the Land For?

Is it going to be residential, agriculture, or a mixture? Are you living in an RV or intending to build a home? Will you live there all year round?

You must ensure your land is zoned for what you intend to use it for. For example, if you want to build a house, it must have residential zoning. When you understand how much space you require, you will find it easier to locate the right piece of land.

Where It Is:

This is one of the more important factors, and there are two things in particular you need to consider:

Accessibility: How easy is it to get to the land? If it will be a full-time home, you need easy access all year round, regardless of the weather. Some factors to consider include:

- Is it near to major highways and roads?
- What condition is the access road in?
- How near is the closest city or town?

If you don't live there full time, you can compromise here, but poor access can make it harder to sell the property in the future.

How Close Are Amenities?

This is also an important consideration, but it depends on what your goals are and if you have dependents. The amenities you might want to be close to include:
- Hospitals
- Doctors
- Schools
- Grocery stores
- Veterinarian
- Farm supply store
- Hardware store

That said, if getting away from everything is your goal, you might want somewhere more isolated. However, the further you are from the amenities, the more difficult it will be to get to them should you need them.

Land Characteristics

Aside from location, you must also consider certain aspects of the land.

Topography:

This is about the physical features, such as slopes, elevation, drainage, etc. These all affect your success, or otherwise, in raising livestock and growing food, not to mention building your house and other buildings. Too steep, and you could struggle to build or to grow crops. If the land floods in places, you may struggle to raise certain livestock or grow crops.

Conversely, it could be the ideal spot if it is well-drained with little to no slope. Terracing is one way of growing crops on gently sloping land, and you won't have to worry about flooding if the soil is well-drained.

Soil Quality:

Soil quality has a considerable effect on livestock and crop health. You need fertile, deep, well-drained soil to allow your crop roots to grow strong. It should be full of the nutrients your plants need, and it must be well-drained so it doesn't flood or become waterlogged.

If possible, get the soil pH and nutrient levels tested. That way, you will know what grows and what doesn't or what you need to do to amend the soil quality.

Legal and Zoning

These are critical considerations as they can affect how you use and develop the land. The two crucial issues are:

Zoning Restrictions:

These regulate how you can use and develop land in certain areas. Before you buy any land, you must find out the zoning laws for the region/state to make sure you can legally homestead on it and build what you need. You can speak to the local zoning department, who will explain the local laws and restrictions. You may also want to talk to a real-estate lawyer to ensure you understand everything fully.

Title Deeds/Ownership:

You must ensure the sellers actually own the title and that no one else has a claim or lien on the land. Get a copy of the title deed and ask a real estate attorney to look at it. They will tell you if any issues could stop you from owning the land.

You may also need to consider title insurance that protects you should any legal issues or claims arise after the purchase. Yes, it is more money, but it is well worth the cost.

Financial Considerations

The financial aspects are perhaps some of the most important, and there are two you need to consider:

Budgeting:

You must plan your budget to determine your spending limits.

https://unsplash.com/photos/a-person-holding-a-bunch-of-money-in-their-hand-9uCmnwC2KR4?utm_content=creditShareLink&utm_medium=referral&utm_source=unsplash

You must set out a budget before you even think about looking for a piece of land. Determine your maximum spending limits on the land, then calculate the costs of property taxes, closing costs, and insurance.

If your chosen land doesn't already have a house, you need to consider the building costs, whether you do it yourself or hire contractors. If it does have a house on it, consider the costs of potential renovations.

If you intend to finance your purchase, consider that some lenders want a bigger down payment than they would for a house. The interest rates may be higher and repayment terms shorter, too. You need to choose the option that fits your circumstances and budget.

Resale Value Potential:

Obviously, this won't be the first thing on your mind, but you may need to consider it in the future. Try to buy in areas with strong demand, as this can increase future prices. Think about:

- Location
- Zoning laws
- Natural resources

These are all important when considering resale value, and unique features, such as lovely views or water access, can boost the value significantly.

Environmental Factors

Environmental factors can get in the way of your ability to work and live, and there are two that you need to consider in particular:

Climate:

Climate affects whether you can raise livestock or grow crops. Some are more suited to specific climates, so research to ensure you can raise the food you intend to raise in that particular climate. Extreme weather also affects things, so if the region is prone to hurricanes, severe flooding, and droughts, not to mention wildfires, you may need to rethink your choice. Some factors to consider are:

- Average temperatures
- Rain/snowfall levels
- How long the growing season is
- Soil type and quality
- Topography

- Elevation
- Microclimates

Disaster Risks:

Natural disasters are bad enough anywhere, but they can significantly threaten a homestead. Consider the disaster risks in the region, as some places are more prone than others. Research the area's history to assess the disaster risk, looking for the following:

- Earthquakes
- Floods
- Hurricanes
- Landslides
- Tornadoes
- Wildfires

You must also consider man-made disasters like chemical spills and industrial accidents. Again, research, look at historical data, and speak to emergency management agencies in the area.

Infrastructure and Utilities

When you purchase land to build a homestead, you must consider the available utilities and infrastructure.

Electricity and Water:

These are critical to a homestead, so think about the following:

- **Water:** Does the land have a reliable water source? A stream, river, or well? If not, how do you gain access to water? Is the water drinkable? Can it be safely used for crops and livestock? You may need to factor in the cost of water tests.
- **Electricity:** Does the land have electricity access? If not, what option do you have? Can you use wind or solar?

Waste Management

Waste management is one of the most critical aspects of keeping your homestead sustainable and healthy. Think about the following:

- **Septic System:** Is the land connected to the municipal sewer system? If not, can you install a septic system? Consider the cost in your calculations and whether the land is suitable for the

installation.
- **Composting:** This is one of the best ways of managing organic waste, so consider if the resources and space are available for a good composting system.
- **Trash Disposal:** Is the land on the municipal trash pickup service? Are you near a recycling plant? You must consider how you will dispose of non-organic waste and consider the cost involved.

Only when you consider all of these aspects can you begin to ensure that your homestead has everything you need to be successful.

Designing Your Homestead

Planning your homestead means ensuring all space can be used effectively, and the right design will mean you don't have to waste time and money redesigning it later. The best time to start is the fall, as the warmer season is ending, and you can get your homestead and gardening areas set up.

First, measure your space and draw it out. Add in all the structures on the land, including measurements, so you know what you can work with. Your drawing doesn't need to be perfect, but it does need to be accurate.

Next, check your local regulations, especially if you want to keep bees or livestock, including respecting the distance from the property lines the law demands.

Determine what you want. Are you after a fruit orchard? Bees? Chickens and ducks? Or something larger, like cattle and horses? Do you need an area for your pets or children? List everything and then work out where it will go on your plan.

Planning Your Layout

Make sure the must-haves are mapped on your design layout first. For example, if you really need a vegetable garden, head out into your land and measure the light and shade at various times during the day. A veggie garden needs maximum sunlight and well-drained, fertile soil. If you live in the north, you should choose an area with the most sun. Conversely, in the south, choose an area that gets shade in the afternoon. Bear in mind that buildings may provide shade.

Make sure bees and livestock are positioned as per regulations. If there aren't any, put them where it suits you, but keep in mind the suitability of the land for the purpose intended. Play around with several layout ideas until you get the one that works for you.

Layout Plans

If you have a suburban plot and are finding it hard to plan your homestead, here are a few ideas to help you.

Large Suburban Plots: these days, if you have a large plot, you likely have a large house but very little land. However, that doesn't mean you can't build a homestead. Careful planning ensures you can fit in a veggie gardening area, be it raised beds, vertical planting or containers, a couple of beehives, or a few chickens, and still have space for your kids to play. Use all the land around the house to full capacity, bearing in mind the laws about distances from neighboring plots.

Medium Suburban Plot: having more space in the backyard means having more space to be self-sufficient, and you can use the front yard, too. Some ideas include:

- A fruit-tree guild in the front yard
- An area for composting
- Beehives
- Small livestock, i.e., chickens and rabbits
- Keyhole garden bed
- Greenhouse
- Small pond

Small Suburban Plot: perhaps you live in a duplex and don't have as much space as you would like, but you can still use raised beds and containers. Some ideas include:

- A perennial veggie bed
- Chicken tractor/mobile coop
- Rabbit hutch with run
- Patio planters

Of course, if you have an acre or more, you can do all this in a much-scaled-up version. It all comes down to using your space in the most effective way possible while respecting local planning laws.

Homestead Planning Tips

Here are a few tips to help you in your planning:

- **Use What's Already There.** If your land has established trees, gather up the dead leaves and make compost, or use them as part of a food forest. If there's a pond, add edible perennials.
- **Study Building Designs.** Work out what will work regarding beehives, hutches, pens, coops, and greenhouses. Use existing buildings where possible and renovate them to meet your needs.
- **Maximize Your Space.** If you are short on space, use trellises and fences for vertical planting, and consider raised beds and keyhole gardens to grow more food in a small space.

Building Your Homestead

You may find it cheaper to build your own home, especially if you can't afford to buy land with a home already on it. The cost of property is very high now, and more people are choosing to build their own, especially when starting a homestead.

Building your own home means building it how you want it, not having to put up with or renovating what's already there. Things to consider are:

Building Materials:

There are plenty of cost-effective building materials that can help you save money and time:

- **Prefab Panels:** These are usually made from concrete, steel, or wood, made off-site, and shipped for assembly. These take less time to build, which means less cost, fewer workers, and less skill. They are often considered eco-friendly because there is less waste, and prefab homes are energy efficient because of their enclosed design.
- **Precast Concrete:** Precast concrete panels are much cheaper than building with concrete and offer insulation, fireproofing, and security.
- **Shipping Containers:** One of the more popular home choices these days, shipping containers are eco-friendly as you are recycling an existing container. Already constructed with a roof, walls, and a frame, these can save money on materials and labor, and it's much quicker to put a home together.

- **Reclaimed Wood:** This is often repurposed from wooden structures, such as boats, shipping crates, and old barns. They are eco-friendly because you aren't cutting down trees for fresh wood. They are also cost-effective and give your home a unique look.
- **Bamboo:** Bamboo is fast-growing and cheaper than wood. It is versatile and sustainable and can be used for roofing, framing, flooring, and fencing. It is also considered stronger than steel.
- **Bricks:** One of the highest-end materials you can use, bricks are also cost-effective, depending on the type you use. Solid bricks are expensive, but building a wood frame and a freestanding brick wall is much cheaper. However, it can be labor intensive, and using a contractor will significantly increase costs, too.
- **Cob:** A natural material made from sand, clay, straw, or another organic, fibrous material, cob is one of the cheapest and most flexible building materials. You can even use it to make bricks and stack them, creating a good foundation to build on.
- **Eco Bricks:** These are made from plastic bottles filled with plastic waste, reducing plastic and pollution in the biosphere and creating a much cheaper building material.

Other things you can do to keep construction costs down include:

- **Easy Shapes:** Don't make your home layout complicated. Keep things simple with squares and rectangles to keep your costs down. If you want a bigger home, go up rather than out – it's still cheaper. You can apply this to the roof, too. The simpler the design, the cheaper the cost.
- **DIY:** If you have construction experience, this can be cheaper than hiring contractors for the whole job, but you must have the right knowledge, skilled labor, and materials.
- **Energy-Efficiency:** Building an eco-efficient home saves pollution and significantly reduces your energy bills.
- **No Expensive Finishes:** These can wait until you've saved up more money. A solid, secure home is more important at this stage.
- **Group Water Storage:** Try to plan so your water requirements are near to each other, such as your bathroom, kitchen, laundry

room, etc. This reduces the need for more plumbing materials and can keep your costs down.

Chapter 2: Sustainable Living Considerations

One of the most important parts of homesteading is being self-sufficient, which means managing your water and energy supplies. You also need to consider reducing your carbon footprint, and you can do that in several ways. This chapter will look at water and alternative energy before moving on to recycling and waste disposal.

Water

Water is critical to everyone, and you'll need it for drinking, cleaning, cooking, bathing, and irrigating your land. Having a reliable source is essential, so here are some ways you can learn to manage your water if you are not on a main supply.

Rainwater Harvesting

Rainwater harvesting allows you to store rainwater for later use.
https://unsplash.com/photos/black-round-metal-tank-on-brown-wooden-table-L7Ps6-zLDpI

This is when you collect rainwater and store it for the many uses you'll need it for. It is one of the most eco-friendly and sustainable practices and is incredibly helpful for homesteaders as it gives them a free water source.

You can do this in several ways, but the more common method is by using gutters, cisterns, or rain barrels to collect water from the roof. Rain barrels come in all sizes, from 50 gallons up, and cisterns are much larger, capable of storing thousands of gallons.

When you set your system up, here are some aspects you should take into consideration:

- Roof type and size
- Amount of rain
- Intended use

The best roof materials are tiles, metal, or asphalt, while asbestos, cedar shakes, and other similar materials come with a contamination risk.

You can use rainwater for cooking, drinking (it must be filtered and treated), irrigation, and cleaning. As it is naturally soft, there are no chemicals or chlorine, so it's ideal for everything.

Wells

If you have a good well on your property, that's great. They are consistent and reliable, tapping into water beneath the ground and supplying a constant source for all your needs.

Wells come in all types, including dug, drilled, and driven. Drilled wells are the most common but require a drilling rig, while dug wells mean getting the digging tools out. A pipe is driven into the ground for a driven well.

Various factors affect depth and yield, including geology, well type, and water table. If building your own, employ a hydrogeologist or professional driller to determine the well location and maximum depth and to ensure it is within your local regulations and follows safety standards.

Getting a continuous supply of water means regular maintenance is required. That includes:

- Water quality tests
- Structure inspections
- Pressure and pump maintenance
- Cleaning the surroundings

Water Conservation

Another important part of homesteading is water conversation. When your resources are limited, using what you have efficiently is critical:

- Mulch your garden to keep moisture in
- Go for drought-tolerant plants
- Use drip irrigation/soaker hoses
- Collect greywater (showers, sinks, dishwashing, laundry) and reuse it in the garden
- Make sure you have no leaks
- Use low-flow faucets/toilets/shower heads to reduce usage
- Use rainwater harvesting systems
- Don't over-water gardens, and only water during cooler hours

Alternative Energy Sources

When you start a homestead, you can use alternative energy sources. One of the biggest reasons for the popularity of alternative sources is that they are eco-friendly, but they also offer a cheaper alternative to mains energy. Here are some of the main sources:

- **Wood Fuel:** Used for heating and/or cooking, wood fuel is incredibly cost-effective compared to electric or gas. While trees take a long time to regrow, wood is sustainable and renewable, and if your land has plenty of trees, you have a great ready-made source.

- **Solar:** Solar is one of the most popular alternative sources, and if done right, you can generate more than enough power for all your appliances. You can mount panels on your roof or set them up on your land somewhere. However, hiring a specialist to install the system is best unless you are experienced, as your panels must connect to your existing wiring. You could also consider a battery storage system to give you constant electricity. If you live in the north, you may want to consider using solar with other alternative sources, as you may not get enough sun to power the system.

- **Wind:** The next most popular wind generation requires wind turbines, which can be noisy. They offer a clean energy source, but you need to live in a region with steady winds. You can also have a battery system to store excess energy. You must check your local regulations, though, as not all areas allow wind turbines, and there will likely be regulations about placement.

- **Water:** Hydropower works 24 hours daily, provided you have the proper setup. A small stream won't cut it. If that's all you have, you'll need to tie it in with other sources.

- **Methane:** A relatively new option, methane is fast becoming popular among homesteaders, particularly those with large numbers of manure-producing livestock. Simply put, it uses gases produced by organic matter as it rots to produce fuel. You can also produce methane with composted plant materials, and it's pretty easy to set up.

Waste Reduction and Recycling

Recycling and waste reduction are hot topics right now, and homesteading is the perfect place to give it a go. Even if your land is on a trash collection route, you still want to consider reducing the amount of waste and reusing/recycling where possible. Here are some ways to help you get started:

Paper/Cardboard

There are so many ways to reuse paper and cardboard that throwing it away is senseless. You can use it in the following ways:

- Reuse toilet roll tubes and egg carts as biodegradable seed starters
- Cut up used printer paper to make notepads
- Weave packing paper (Amazon, etc.) into baskets
- Reuse old wrapping paper as photo matting
- Use newspapers in the garden as mulch. Tear it up and layer it around your plants, and it will eventually rot down, feeding the soil.
- Shred paper to use as chicken bedding
- Chuck it in the compost bin
- Use it to start your wood fire

Plastic

Everything is made of plastic these days, and it's one of the worst materials as it doesn't biodegrade. However, it does break down into small pieces, causing damage and death to wildlife. Here's how to reuse plastics on your homestead or reduce your usage altogether:

- Reuse plastic shopping bags as pillow stuffing
- Use large plastic bottles as mini greenhouses in the garden
- Ziploc bags can be reused many times
- Instead of plastic shopping bags, buy reusable ones
- Buy drinks in reusable bottles
- Don't use disposable straws and cups. Buy/use reusable ones instead

Glass

Glass is one of the easiest materials to reuse and is environmentally friendly.

- Turn glass bottles into planters.
- Reuse glass jars and bottles as storage vessels, especially when you buy things like flour, sugar, beans, etc., in bulk.
- Reuse broken glass as decorations on the wall or in art projects.

Food Waste

One of the best ways to reuse or recycle is to compost what you can. That applies to food waste, as a composter can turn it into a fantastic fertilizer. Composting breaks materials down into nutrient-rich compost, reducing reliance on landfills and fossil fuels and reducing emissions.

There are a few types of composting to consider:
- Aerated (turned) windrow composting
- Aerated static pile composting
- In-vessel composting
- On-site composting
- Vermicomposting

Which one you use will depend on your climate, homestead size, use, and how much compost you need. You could even run two or three systems separately. You can make a composting bin by using a large bin with holes in it or building one outdoors out of old pallets and wire netting. Alternatively, buy a purpose-made one.

How to Reduce Your Carbon Footprint

When you start a homestead, there's every chance you want to try living off the land, which is great for your finances and the environment. When you rely less on external entities, you get more control over your interaction with your land, and you may even be looking for more ways to reduce your carbon footprint. Here are some for you to consider:

1. **Make Your Own Clothes or Buy Second-Hand:** While it may be fun to have new clothes, it takes numerous resources to make them. For example, cotton is incredibly water-intensive, while man-made fibers use coal-based methods to make them. Save money by buying clothes from second-hand stores, buying material and making your own, or altering your clothes.

2. **Compost Food Waste:** As mentioned above, composting is low-emission, creating wonderful compost and fertilizer for your veggie garden. Make piles of organic waste, such as plant material, leaves, and food scraps, interspersed with layers of paper and cardboard.

3. **Use Renewable Energy:** Moving off the grid is a great way to reduce your carbon footprint. Think about using any of the renewable sources mentioned in the earlier section. You could also consider purchasing a solar generator to start with to see how you get on before you go full solar or use wind turbines if your land is suitable.
4. **Grow Your Own Food:** Raising livestock and growing fruit and vegetables is a great way to reduce your reliance on mass-produced foods at the grocery store, with the benefit that you know exactly what has gone into your food. By producing your own, you are not contributing to the fuel pollution to deliver your foods to the markets, not purchasing inorganic foods smothered in chemicals and full of growth hormones, and there is no waste, either. You could consider joining a farmer's community, where you can exchange foods or purchase organic.
5. **Recycle and Reuse:** Wherever possible, reuse or recycle everything. If you have paper or cardboard that you can't use in another way, chuck it in the compost bin. Reuse egg cartons as plant starters, reuse glass jars and bottles for storage, and so on.
6. **Collect Water:** Set up rainwater systems to collect water you can use around your property. You can use it just for watering the garden or go the whole hog and set up a whole-home system.

Just a few simple steps can help reduce your carbon footprint significantly, and you'll learn some interesting uses for things you never thought of. At the end of the day, everyone has to do their bit to help save the planet, and even the smallest things can help.

Chapter 3: Should You Go Off-Grid?

Going off-grid means not relying on any public utilities or the rest of society for your everyday living. It involves being self-sufficient to the extent that your energy is supplied by renewable means, you have your own water source, and you grow and raise all your own food.

At one time, living off-grid meant living in a remote rural area, well away from everyone and everything. These days, you can live anywhere and be off-grid, at least partly, but if that's what you want to do, there are some things you must consider.

First, decide where you want to live. The best place would be a large plot of land with a water supply, preferably running water but not a main supply. You'll need some kind of water collection system, too.

Vegetable gardens are a necessity when it comes to living off-grid.
https://unsplash.com/fr/photos/plante-verte-et-rouge-sur-cloture-en-bois-blanc-hvSBya7hX2Q

You'll need enough land to create a decent vegetable garden, have fruit trees and bushes, and raise a certain amount of livestock for food.

In terms of power, you'll need to decide on solar, methane, wind turbine, or some other renewable source, but factor the costs of the system into your budget. You'll also need to consider a battery storage system if you choose wind or solar. Otherwise, you'll only have power during daylight hours.

Lastly, you need to consider a waste disposal/septic system and add the costs to your budget.

Sustainability of Off-Grid Living

Off-grid living is more sustainable than relying on standard means of energy, water, and food supplies, partly because you won't use as much energy but mostly because off-grid systems are renewable.

20% of greenhouse gas emissions come from powering, cooling, and heating homes, just in the USA. When you live off-grid, you do your bit for climate change by not adding to those emissions. On top of that, off-grid homes typically have better insulation, don't use so much energy, and use renewable energy sources instead of fossil fuels.

You no longer rely on grocery stores because you grow and produce your food, including eggs, milk, meat, and vegetables. You no longer consume anything that is not organic and no longer eat chemical-laden foods wrapped in plastic and transported, sometimes a long way. And if you have your own compost heap, you save water and soil and reduce your wastage.

Pros and Cons

Off-grid living comes with advantages and disadvantages.

Pros:
- **You Save Money.** The upfront costs of setting up can be expensive, but your costs go down after the initial investment. Your energy bills are cheaper, and you purchase less food as you grow your own.
- **Connection with Nature.** When you live on a large piece of land, you get to immerse yourself in nature and escape modern life and all the stresses it brings. It's long since been recognized that just 2 hours a week spent outdoors in nature significantly improves mental and physical health. Off-grid living means being

surrounded by nature 24/7.
- **Sustainable and Self-Sufficient Lifestyle.** By choosing renewable energy and growing your own food, you learn to rely on yourself and not contribute anywhere near as much to society's throwaway culture.

While off-grid living is rewarding, it can also be incredibly challenging, not to mention requiring commitment.

Cons:
- **Substantial Investment.** Setting up an off-grid homestead is not cheap, as you have to consider land purchase, licenses, permits, building costs, water supply, renewable energy, and waste disposal systems.
- **Hard Work.** You definitely need some skills, or you need the money to hire people who know what they are doing to set up your energy and water systems. You also need to know how to raise livestock, possibly butcher it yourself, and learn how to grow vegetables and turn everything into food to survive.
- **Things Can Go Wrong.** Nothing ever goes smoothly, and you must expect things to go wrong, such as your water or power supply failing at times.
- **It Can Be Isolating.** If you choose to buy a piece of land in the middle of nowhere, be aware that you may be quite isolated and not see another person for days, weeks, or even months unless you head into the nearest town or city. Ask yourself if that's what you really want.

Chapter 4: Building Your Homestead

When you've purchased your land and are ready to begin your new life, the first thing you need is a roof over your head. No doubt you've been told you should spend a year on your land before you build a home, but most people don't fancy camping for that long.

Few homesteaders have the necessary experience or skills to design or build a home, but the problem is likely because you're thinking too big. Your first home doesn't have to be massive. Think small. Think simple to start with because you can always expand later on. All you need are a few basic necessities to get you going.

First, you need a source of running water, be it a well, mains, or a spring. Secondly, you need a power source of some kind. Once you've got that, you can plan out a basic home that will serve your purpose until you can expand into something more luxurious.

The Bare Necessities

Here are the essential things you need in your house:

1. **Somewhere to Sleep**

A sofa bed can serve as a bed at night and a sofa during the day.
https://unsplash.com/photos/white-bed-linen-near-window-tGo2ngNyKyM?utm_content=creditShareLink&utm_medium=referral&utm_source=unsplash

While having a room specifically for sleeping, if there's only you/your partner, a comfy bed tucked away or a sofa bed will suffice. Provided you've got decent warm bedding, you'll be just fine. You can even use a bed as a sofa during the day by switching pillows for cushions. If you need a bit of privacy, a curtain rod and curtains will work.

2. **Somewhere to Store and Make Food**

Again, having a separate kitchen is nice, but it isn't necessary to start with. All you need is somewhere flat and solid to work on near your stove. A solid table will work, or a solid board on heavy brackets. You should invest in some kitchen cabinets, too. They don't have to be new ones, either. Pick up some old ones and upcycle them. A couple of shelves above the worktop and a basic sink, and you're all set. You can drain the kitchen sink into a container outside and use it to water your garden. You also need a couple of basic baking dishes, crockery, and cutlery, and for closed storage, a few kitchen cabinets stacked up will work well.

3. **Heating and Cooking**

You will need some kind of stove, but if power is an issue, use a wood stove or propane gas. A wood stove will also provide heating. If you do have power, invest in a microwave, crockpot, toaster, kettle, and a basic two-ring hob. That will get you started, and you'll at least be able to cook something.

4. **Somewhere to Store Your Clothes**

If you are switching city life for a homestead, there's a good chance you've downsized your wardrobe to just the bare essentials. Use storage bins or drawers beneath the bed if you don't need to hang anything up. If not, get a chest of drawers. Again, think second-hand that you can upcycle. This will also double as a sideboard and not take up much floor space.

5. **Bathing and Toilet**

If you live somewhere with warm weather all year, an outdoor shower and outhouse will do the trick. However, if your region gets cold, you'll want somewhere indoors and a little warmer. It doesn't have to be big, just enough for a basic toilet (think sawdust or compost for now) and a washstand, and it can double as a wet room for a basic shower. Again, link the drain to water storage outside so you can reuse the water.

6. **Somewhere to Store Your Tools and Work**

You'll need tools for your homestead, so you'll need somewhere to store them. Add a toolshed onto the side of your house and give it an entrance from the outside and one into the house. Make sure it is secure. If you have space, add a covered workspace, too. This will keep you cool in warm weather and sheltered from the elements in cold weather. You can also use it to hang your washing out and hang the washing out. Don't be mean with the space. Ensure it is at least 8 feet wide and as long as needed so you have space to work.

Let's Get Building

Those are what you need as a minimum, so get drawing and plan your house. The rest of this chapter will give you some building instructions and help you get your first shelter built. The building instructions will be much the same for most plans. You will now learn how to build a deck on posts and add walls with conventional framing between each post. Those posts provide support for the roof rafters.

Step One:

a. Mark out your building plan on the ground. You can use string and poles to do this, ensuring you mark the position of the corners and every post going into the construction.

b. Dig the post holes deep enough that they go beneath the frost line or until you hit a big rock. If the rock isn't deep enough in the ground, you may need to rethink the position. Three feet is usually deep enough to support a 4 x 4, 14-foot post.

Step Two:

a. Set your posts in place and concrete them in. Make sure the posts have been treated first. If you are building a single-story building, 4 x 4 posts are sufficient. Choose 5 x 5 if you are building a two-story home or your ceilings will be high. Be aware that you will need help to handle them as they are heavy and awkward.

Step Three:

a. Use 16-penny nails or lag screws to attach the rim joists. Make sure your floor joists are a minimum of 2 x 10. These will span 14 feet without needing support beams, but feel free to add supports if you want them.

b. Use standard joist hangers to hang the floor joists on 2-foot centers. You can use offcuts from your 2 x 10 joists as blocks between the floor joists, but make sure they are staggered every 6 to 8 feet. This stops the joists from bouncing or twisting.

Step Four:

a. Use plywood for the subfloor and lay it out. Nail or glue it so it is secure. If you prefer to use planks, ensure the joints are staggered and nail them down.

Step Five:

a. Lay the sills out, marking them on 16-inch centers for the studs. Make sure the wall sections are framed to fit between the posts and frame the windows and doors.

b. Use plywood or sheet siding as exterior wall sheathing and go over the doors and windows. You can cut those out later.

Step Five:

a. Go with a simple roof structure, a single slope similar to a shed roof. Using 16-inch centers, mark out where the rafters will go.

However, if you are not using roof insulation, you should go with 2-inch centers. Secure the first and last rafters in place with nails, making sure you have the right-sized overhang.

b. Secure the fascia boards at each end with nails and use them to support the rest of the rafters.

c. If you are using roof shingles, place plywood or boards over the shingles and add black tar paper. Nail on a metal drip edge and place the shingles over the entire roof.

d. You must nail purlins across all the rafters if you choose metal roofing sheets. Then, lay the metal roofing over the top and secure them down.

e. Add soffits beneath the overhang if you need the house winterized or don't want insects getting in.

f. If you want a gable roof, buying ready-made trusses is best, as they are easier to install and cheaper.

Step Six:

a. Install your doors and windows.

b. Add siding if that's what you've opted for, and make sure the entire building is caulked in.

Step Seven:

a. Now is the time to wire your house for electricity, but keep things simple. It's probably best to hire an electrical contractor unless you already know what to do. You can get away with a single circuit for the lights and a single circuit for the kitchen if you don't have large appliances, but splitting the kitchen plugs between two circuits is recommended. That way, you can extend the circuits to other areas.

b. If you want a large range cooker at some point, you will need a 220 circuit only for that and a separate one for a refrigerator. It's best to plan for the future even if you only built a basic house for now.

c. Suppose you are hooking up to the main supply. In that case, the power company will do the connection for you, or if you are going full solar, the installation company can do it.

d. You should also do the plumbing now. Hopefully, you have placed your bathroom and kitchen on the same wall, so the plumbing is all in one place.

Step Eight:
a. Install the wall insulation if you are using it. Use fiberglass batts as they are cheaper and easy to install.
b. If you don't want insulation, you have two choices. Leave the walls open to the framework or install a wall surface. Dry-walling is the easiest and most cost-effective way and can easily be painted over. Alternatively, use pine boarding.
c. You can set up your furniture and furnishings when the walls are up.

Layout Ideas

Layout One:

A small, simple design would be one for one or two people. At 368 sq. feet, it would have a main living area with a 1o-foot kitchen counter, open shelving, and a small sink. The sleeping quarters are a basic platform, big enough for a queen-sized mattress, and built-in drawers beneath.

A small room takes care of a shower/toilet area, while a wood stove provides the heat and cooker. The main wall has double French doors that lead to a large porch, and a small shed at the back works as tool storage.

You can add loft space over the whole area or just part of it, and it has a gable roof and plenty of windows.

Layout Two:

This is a larger space, 672 sq. feet, including a barn and work area. There are two bedrooms, one small, one larger, and high ceilings. The smaller room is enough for a single or bunk bed with storage space beneath, while the larger room has a queen-sized bed platform with storage beneath and sufficient space for small furniture, such as a chest of drawers.

The living area and kitchen are in the center of the layout, with a 10-ft kitchen work surface and an area for a wood fire. The bathroom is a 6 x 7 foot space, while the storage room is 6 x 10 foot. The barn has enough space for a work area, a chicken house, and one stall for a goat, sheep, or cow.

It isn't huge but more than enough for a new homesteader and can easily be extended if needed. Loft space could also be installed.

Layout Three:

This is much larger and consists of a pair of large structures with shed roofs. They are the same size and face each other with a large 20-foot courtyard in between. The courtyard can be left open or covered.

There are 3 bedrooms in one structure: 2 smaller ones with space for single or bunk beds with storage space beneath and a larger one with space for a double bed. The second structure has a large living area and combined kitchen, with a storage area and toilet room.

The courtyard can be fenced in or left open, or it can have trees for shade, whatever you want to do with it.

If you build one of these and choose to build a larger, more permanent structure later, this layout can be converted into a large barn. The building fronts facing the courtyard are high-sided and can easily have trusses between them, supporting a roof. The bedrooms could be turned into animal stalls, while the kitchen/living area could become a work area or somewhere to store food. Built right, the roof above the existing courtyard could be turned into a hayloft, and doors and walls can be used to enclose the space.

All homesteaders want somewhere safe, warm, and dry for themselves and their families, and that is easily accomplished if you keep things simple, at least to start with. And once you are settled, you can either expand your current house or build a new, larger one and repurpose the old one.

Section Two:
Living Off the Land I: Plant-Based Food

Chapter 5: Getting Started with Gardening Edibles

Before you move on to this chapter, you should set up your homestead and be ready to start growing your own fruit and vegetables.

Starting a Vegetable Garden

Growing a vegetable garden allows you to be self-sufficient.
https://www.pexels.com/photo/person-checking-his-vegetable-plants-7658823/

If you've never had the space for a vegetable garden before, you should follow a few loose rules to get the best results.

1. **Choose Somewhere Sunny:** Take a good look around your land. Vegetables need sunlight to grow, so choose somewhere that gets at least 6 to 8 hours of sun daily.
2. **Make Sure It Is Safe:** It's no good planting vegetables on land that floods when it rains or is prone to strong wind and there are no windbreaks (shrubs, trees, etc.) Make sure the area is safe first.
3. **Know the First and Last Frosts:** These are important but so often overlooked. Knowing the last frost date in the spring ensures you don't plant heat-loving crops, such as peppers and tomatoes, outside too early. You also need to ensure what you plant can be harvested before the first frost date or is cold-tolerant, like cabbages, leeks, spinach, etc. You also need to be prepared for early and late frosts and have a plan in place to protect your plants.
4. **Start Small:** So many people get excited when they plant a vegetable garden for the first time that they go overboard and plant a huge garden. Don't do it. Gardens are hard work, so only make your first garden a small one that serves the people you are feeding. Start with a plot about 16 x 10 feet, which will provide enough food for an average family for 4 harvests during the summer, with a little left over to store for winter. You can always expand later.
5. **Choose Easy Vegetables:** Only grow what you know you will eat and that you can grow easily. Check to see what grows well in your region – the local garden center or other seasoned homesteaders will know. Some of the easiest plants to grow are:
    - Beans
    - Beets
    - Carrots
    - Cucumbers
    - Lettuce
    - Peppers
    - Radishes
    - Spinach
    - Squash
    - Swiss chard
    - Tomatoes

- Zucchini
6. **Plan Your Garden Space:** The fun part is planning where everything will grow. Draw your garden space on paper and divide it into beds with pathways between them. Plan to grow tall plants at the northern edges, ensuring they don't throw shade over low-growing plants. If you don't have enough space for a big garden, try square-foot gardening. Planned right, it's surprising how much you can grow in a small space.
7. **Prepare the Soil:** There are two basic methods you can use to prepare your soil:
    - **No-Dig Gardening:** Place cardboard or newspaper over the garden space and top it off with a thick compost layer. Your plants are planted straight into the compost, meaning you don't need to dig or amend your soil.
    - **Traditional Gardening:** This requires you to till or dig the ground, but make sure it is dry first. If you dig wet soil, it gets compacted, and your plants will struggle to grow. Till the garden, then rake it over and till it again. Once the soil is loosened off, add well-rotted manure or compost and till it again. This should be done in the fall. You can hand dig it if you don't have access to a tiller, but this is a lot of work. Dig it over, add manure or compost, and rake it smooth.
8. **Start Planting:** If planting seeds directly into the ground, follow the package instructions for spacing. Cover them with soil, firm them down, and water. You want to keep the soil moist but not overly wet until the plants have started growing and are established. If you start with seedlings, harden them off before planting. This means placing them outdoors during the day and bringing them in overnight, gradually increasing how long you leave them outdoors.
9. **Add Mulch:** This ensures the soil retains moisture, stops dirt splashing onto your plants when it rains, and keeps the weeds down, all saving you a lot of work.
10. **Watering:** Watering is the most crucial part of growing veggies, and your garden will need around an inch of water per week. If you get plenty of rain, you don't need to worry, but you will definitely need to get the hose out in dry areas. Soaker hoses are the most efficient way of watering as the water goes right to the

roots and nowhere else.

As you can see, starting a vegetable garden isn't too hard, but don't underestimate the work involved. A well-planned, properly located garden can feed you and your family all year if you do it right.

Chapter 6: Harvesting and Making Your Own Plant-Based Food

The more you work in your garden, the more you will learn. This is very true, especially when growing fruit and vegetables. Most seed packets give full planting instructions and also give you an idea of how long the plant takes to mature. However, those are only guidelines because many factors can influence harvest time.

Knowing the "official" number of days a plant takes to mature is a good starting place, but you also need to use common sense to know when your vegetables are ready to pick. The first part of this chapter will guide you through harvesting, while the second part provides a few recipes to help you get started using your harvest.

Harvesting Principles

There are principals you should follow when it comes to harvesting.
https://unsplash.com/photos/yellow-and-red-tomatoes-on-green-plastic-crate-hmoDcZnB7uw

These principles give you an idea of how and when you should harvest:

- **Peak Flavor and Nutrition:** Some vegetables don't need to be entirely ripe to be at their best. Summer squash, turnips, beans, and peas should be harvested when they're tender and young. On the other hand, those that should be ripe for full flavor include melons, winter squash, peppers, and tomatoes.
- **Size:** This is a good indicator, but knowing when vegetables are ready to pick takes practice. Check the seed packets for a good indication because all varieties differ.
- **Pick Often:** Many gardeners make the mistake of not harvesting regularly. If you leave your beans on the plant, they can quickly turn tough, while a ripe but small zucchini can quickly expand and become overripe in just a couple of days. Plants have one goal, which is to reproduce. By picking often, you keep that plant producing plenty of food for you. Stop picking, and the plant stops producing.
- **Use the Correct Tools:** A number of vegetables can be finger-picked, including peas, lettuce, and kale, but you'll find many more that need to be cut using scissors or cutters. Finally, forks

are best when it comes to root crops.
- **The Ideal Conditions:** Vegetables are at their best when they are harvested, but that quality quickly declines. When the dew has dried and the crops are at their juiciest and sweetest, early morning is the ideal time to harvest. Harvesting during the hottest part of the day is not recommended, as leafy crops such as kale and lettuce can wilt quickly.
- **Harvest with Care:** Cucumbers and tomatoes are examples of vining plants that require appropriate trellising to prevent breakage of the plant stems. Avoid tearing the veggies off the plant, as this might harm it and make it more susceptible to disease. Additionally, you should avoid working in your vegetable patch during rainy seasons since you may inadvertently transfer fungi and diseases from one plant to another.
- **Outer Leaves First:** Some crops grow from the middle outwards – think cabbage, lettuce, and similar leafy veggies. Unless you are picking the whole plant, start with the outer leaves, leaving the center to continue growing.

Harvesting Guidelines

This is not a complete list, but some of the more common vegetables you are likely to grow:

Arugula:
- Pinch or cut leaves when 2 to 3 inches. Young leaves taste best, but you can eat older leaves until the plant bolts.

Asparagus:
- Harvest when 6 to 8 inches tall with tightly closed tips.
- Break them by bending or cutting them off below or at soil level. Do not cut any nearby spears that haven't come through yet.
- Continue harvesting for 4 to 8 weeks and stop once the spears are thinner with open tips.
- Be aware that an asparagus bed can take at least 3 years to develop before you can harvest it.

Beans – Snap:
- Bush and pole beans are harvested when they are about pencil-thick, with small beans inside.

- Do not harvest immature, thin pods, as they won't have much flavor.
- Avoid harvesting heavy, protruding pods since the beans within are bland and rough. Harvest them as dry beans by leaving them on the plant.

Beans – Dry:
- When the pods turn yellow and dry out in the fall, mature pole or bush beans are harvested.
- Snip them from the vine and spread them in a warm, dry place to dry for 2 to 4 weeks.
- When 100% dry, shell the beans – they should be hard and shiny – and store them.

Beets:
- Harvest beets at about 1 ½ to 2 ½ inches in diameter (Any larger, and they may be woody with less flavor)
- Beets must be harvested before the ground freezes

Broccoli:
- Harvest when the head is 3 to 6 inches in diameter, with dark green, closed buds.
- Cut it about 6 inches below the head, and smaller heads will grow off the side.

Brussels Sprouts: Harvest from the bottom when the sprouts are 1 to 1 ½ inches in diameter, twisting or cutting them off.

Cabbage: Harvest when the cabbage head is about football-sized or bigger and solid. Any longer, and the cabbages will split, expanding in size.

Carrots:
- Most varieties can be harvested at about an inch round. Pull one up to check the size.
- Spring-planted carrots should be harvested as soon as they mature, or they will become bitter.
- Fall-planted carrots can stay in beds over the winter so long as they are covered by straw or grown in a polytunnel.

Cauliflower:
- Keep an eye on the maturity date indicated on the seed packet, as cauliflower can rapidly transition from its peak to beyond it.

- The head should be regular-shaped with tight curds and not yellow. Left too long, the plants will bolt.
- Cut just below the head, leaving some leaves on.

Collards:
- Harvest leaves when they are 6 to 8 inches long and tender, as older leaves are tough.
- Start with the large leaves at the bottom and leave smaller leaves to keep growing.

Corn:
- You can test the corn for ripeness when the silks turn brown and dry. Feel the tip through the husk. It is ready to pick if rounded but not if it tapers and is thinner at the top.
- Alternatively, use a fingernail to prick a kernel. If the liquid is milky, the corn is ready.
- Harvest first thing in the morning, as the corn is sweeter.

Cucumbers:
- Pickling cucumbers should be cut when 2 to 6 inches long.
- Slicing cucumbers should be cut when 6 to 8 inches long.
- In all cases, they should have dark green, glossy skin. If yellow or dull, the cucumbers will be bitter and tasteless.

Eggplants:
- Eggplants must be cut with a little stem left on.
- Wait until they are about half the recommended mature size and have a glossy, all-over color.
- They are overripe if they are dull and soft.

Garlic:
- You can harvest your garlic when the tops start to dry off.
- Make sure you aren't bruising the bulb as you are lifting the plant out.
- Clean the soil off with water and leave them to dry. Make sure they're placed on a net tray and in a dry shade that is warm for approximately two to three weeks. Make sure it has good air circulation.
- When they are dry, cut off the tops and store them.

Kale:
- Leaves can be harvested as needed when 6 to 8 inches long.
- Start with the outer leaves and cut them off at the base.
- Baby leaves can be harvested for salads when they are 2 to 3 inches long.

Leeks:
- Harvest when the leek is an inch thick.
- Pull the entire plant from the soil.

Lettuce:
- Lettuce heads can be harvested when they are medium-firm, feel full, and are approximately 6 inches round.
- Leaf lettuces can be harvested when the leaves are 3 to 4 inches tall.
- They will produce new leaves, so continue harvesting regularly until the plant produces a stem in the center. This indicates the lettuce is about to bolt.

Melons:
- Cantaloupes should be harvested when the fruit separates easily from the stem, they have a sweet smell, and the rind is yellow. Harvest as soon as the dew has dried in the morning.
- Honeydews should be harvested when the rind is yellow or white all over. They will need to be cut from the stem and should be harvested once the dew has dried.
- Watermelons are ready when the underside turns cream or yellow, and the topside is a dull color. They must be cut off the plant, leaving a couple of inches of stem attached.

Onions:
- Green onions (scallions) are ready to harvest when the tops are 6 inches tall and can be pulled out whole.
- Bulb onions are harvested when the bulbs are about 1 to 2 inches round, and the tops are brown and have fallen over. Pull the whole plant, and dry the onions by spreading them out or hanging them somewhere warm, dry, and ventilated for a couple of weeks. Then, brush the soil off, cut off the tops, and trim the roots before storing.

Parsnips:
- Allow a few light frosts before you harvest parsnips as it improves the taste, or leave them in-ground, covered in straw, for the winter, harvesting as needed.
- Ensure you harvest before the spring, as they lose their flavor if you allow new growth to shoot from them.
- Pull the entire plant, trim the foliage to about ¼-inch, and store in a root cellar or refrigerator.

Peas:
- Snow peas can be harvested when the pods are a mature length but before the peas have filled out the pod.
- Sugar snap peas can be harvested when the pods and peas are plump.
- Shelling peas are harvested when the pods are plump, green, and firm and the peas have grown.

Peppers:
- Sweet peppers can be harvested unripe or ripe. Cut them off so you don't damage the plant. You can leave the peppers to ripen – yellow, orange, and red – to get the full sweetness, but you can also harvest them green. If you do, harvest regularly so the plant produces more. You can always leave later peppers to ripen fully, giving you the best of both worlds.
- Hot peppers can again be harvested ripe or unripe by cutting them off. Red peppers are hotter than green. Make sure you wear gloves, as they can burn your skin.

Potatoes:
- When you harvest potatoes depends on their variety.
- New potatoes can be harvested 6 to 8 weeks after planting, just after flowering. You can harvest a few potatoes from the plant by digging gently into the soil and leaving the rest to mature.
- Main crop potatoes are harvested about 2 weeks after the tops have died. Dig them out carefully so you don't bruise or damage any. If you do, they must be eaten quickly because they cannot be stored.
- Leave the potatoes on top of the soil to cure for a couple of weeks (unless the weather is wet), and store good potatoes

somewhere dark, cool, and frost-free.

Pumpkins:
- The pumpkin should be a uniform, deep color, and the stem should start to dry before harvesting.
- Cut them off the vines, leaving a small amount of stem attached, and spread them out to dry for 10 days. They must be somewhere warm, dry, and ventilated.
- They can be stored in a dry, cool, ventilated area.

Radishes:
- Spring radishes are harvested when small and mild. They get woody and bitter if you leave them in the ground too long.

Shallots:
- Shallots are mature when the bulbs are 1 to 1 ½ inches round, and the tops are brown.
- Cut off the tops, let the shallots dry for a week in a warm place, and then store them.

Spinach:
- Harvest spinach leaves when they are 3 inches long, starting with the outer leaves.
- Continue to harvest as needed until a center stem appears, indicating the plant is starting to bolt.

Squash:
- Summer squash should be harvested when small for good flavor. Harvest after the dew has dried and cut them from the vine. Check your plants and harvest daily, as they can quickly go past their best.
- Winter squash is ready when the fruit is colored all over and the vines are dying off. Cut from the plant, leaving a bit of stem attached, and cure them somewhere warm for 10 days before storing.

Sweet Potatoes:
- These are ready to harvest when the foliage has turned yellow.
- Harvest before the first frost unless grown in a polytunnel.
- Lift the tuners gently from the ground, taking care not to bruise or damage them.

- Cure for one day in the sun, then in the shade for another week or so.

Tomatoes:
- Harvest when they are ripe but firm.
- Harvest before the first frost. Check daily and harvest ripened fruit, as this will encourage more growth.

Turnips:
- Harvest when they are about 2 to 3 inches in diameter for a sweet taste.
- Harvest before the first frosts.

Plant-Based Recipes

When you harvest your fruit and vegetables, you should use them quickly, store them for winter, or turn them into something else. The remainder of this chapter provides some recipes to help you use some of your harvest.

Strawberry Jam

You can also use blackberries, blueberries, or raspberries in this recipe. The lemon peel provides natural pectin to help thicken up the jam. You can use any sugar – granulated, coconut, maple, or agave syrup, whatever takes your fancy, but try to keep all your ingredients organic.

You can add chia seeds and agar agar if you want an even thicker jam.

Ingredients:
- 2.2 lb. strawberries (or other fruit of your choice).
- 8.8 oz. sugar
- 1 whole lemon
- 1 tsp. chia seeds – OPTIONAL
- 6 tsp. agar – OPTIONAL

Instructions:
1. Rinse and hull the strawberries and chop each one into four parts. Put them in a bowl, add thin lemon peel strips, and squeeze the lemon juice over the top.
2. Sprinkle with the sugar and mix until it dissolves. Marinate for at least an hour, preferably overnight.
3. Add to a pan, remove the peel, and bring to a boil. Simmer for 15 minutes, stirring frequently.

4. Using an immersion blender, blend the mixture for a couple of seconds – skip if you prefer chunky jam.
5. OPTIONAL – add the chia and agar to make a thicker jam, which is recommended if you want to use the jam in cakes or pies. Then, boil it for 5 minutes so the agar is activated.
6. Transfer to jars and leave to cool before storing in a refrigerator. Use within 7 days.

Roasted Cherry Tomato Sauce
Ingredients:
- 6 cups cherry tomatoes.
- 1 onion
- 3 cloves garlic OR 3 tbsp. minced garlic.
- 3 tbsp. olive oil
- ½ tsp. salt
- ½ tsp. pepper

Instructions:
1. Preheat your oven to 400°F and place greaseproof paper on a baking sheet.
2. Chop the onion into quarters and lay it on the baking sheet in a single layer with the tomatoes and garlic.
3. Drizzle oil over, season with salt and pepper, and stir to coat the veggies in oil.
4. Bake until the tomatoes begin bursting and brown off a little on top, about 45 to 50 minutes.
5. Place everything in a blender and blitz to a creamy sauce, about 2 to 3 minutes.
6. Cool and store in the refrigerator until needed.

Canned Zucchini Spread
Ingredients:
- 5 loosely packed cups of chopped onions
- 8 1/3 cups chopped carrots
- 6 2/3 cup chopped zucchini
- 2 chopped celery stalks
- 2 ½ cups chopped bell peppers

- 2 chopped jalapenos
- 5 loosely packed cups of chopped tomatoes
- 2 large cloves of garlic
- 1 to 3 tbsp. salt
- 1 tsp. black pepper
- 1 cup apple cider or white vinegar
- Fresh chopped parsley, dill, and cilantro to taste
- Oil for frying

Instructions:
1. Use water or oil to sauté the onions, then add the carrots and stir. Crush the garlic, add that to the pan, and then add the celery and zucchini.
2. Stir, cooking until the veggies have softened, then add the tomatoes, peppers, and vinegar. Simmer for 10 minutes if you are using a pressure canner or up to an hour if not.
3. Add the spices and herbs, cook for a few more minutes, then process in the pressure canner at 11 pounds pressure for 20 minutes for pint jars and 25 minutes for quarts. Don't forget to adjust the pressure for altitude if needed.

Sprouted Bread

Ingredients:
- 1 cup warm wate
- 1 ½ tsp. dry yeast
- 2 tbsp. organic honey
- 2 ¼ cups sprouted wheat flour.
- ¼ cup oat bran *
- 1 tbsp. extra-virgin olive oil
- 1 tsp. salt

* If oat bran isn't available, you can put ¼ cup of quick oats in a blender or processor and process it into fine-milled flour.

Instructions:
1. Add the honey to water, stir to dissolve it, and then add the yeast. Leave it until it bubbles.

2. Put the oat bran, wheat flour, and salt in a bowl and whisk gently to combine. Add the foamy yeast to the dry mixture, add the oil, and stir to combine thoroughly. Leave it for 10 minutes.
3. Lightly flour a surface and tip the dough onto it. Knead it for five minutes, then place it in an oiled bowl and cover it with a clean cloth. Leave it for an hour somewhere warm until it doubles in size.
4. Oil a standard loaf pan, form the dough gently into a log, and place it into the pan. Leave until it has risen to the top edge of the pan, about an hour again.
5. About 10 minutes before the time is up, preheat your oven to 350°F.
6. Bake the loaf for about 30 to 35 minutes. It should sound hollow when you tap it and be golden brown. If it browns too quickly, cover it with a foil tent.
7. Once cooked, leave it to cool for about 5 minutes, then transfer it onto a cooling rack. Cool completely before slicing.

Baby Puree

Making organic baby puree with your harvest is pretty simple. All you need are the veggies and a blender. The following recipe uses 6 sweet potatoes, but you can use any of the following with the same recipe:

- Beets
- Butternut squash
- Carrots
- Green beans
- Greens
- Parsnips
- Peas
- Turnips

Instructions:
1. Peel your chosen vegetables and chop them into chunks.
2. If you have an electric or stovetop steamer, place them in the basket. Add about an inch of water to the pan (do not add the basket yet) and bring it to a simmer.

3. Add the veggie basket and cook, covered until soft. Alternatively, cook them in a microwave.
4. Place the cooked vegetables in a blender and pulse to a smooth consistency. If necessary, add a little of the water you used for steaming.
5. Allow it to cool and refrigerate until needed, or divide it into portions and freeze.

You can also use fruit, such as:
- Apples
- Blackberries
- Blueberries
- Mango
- Peaches
- Pears
- Pineapple
- Strawberries

Spicy Plum and Apple Chutney
Ingredients:
- ½ garlic bulb
- Thumb-sized chunk of fresh ginger
- 1 large onion
- 1 lb. 2 oz. apples
- ½ tsp. cumin seed
- 1 stick cinnamon
- 1 cup apple cider vinegar
- 1/2 of 1 bulb of garlic
- A small thumb-size piece of ginger
- 1 large onion
- 1 pound and 2 ounces of apples
- 1 lb. 2 oz. plums
- 1/2 teaspoon cumin seeds
- 1 cinnamon stick
- 1 cup apple cider vinegar

- 1 cup golden caster sugar
- ½ tsp. salt

Instructions:
1. Sterilize your mason jars for 30 minutes at 300°F.
2. Peel and sliver the garlic, shred the ginger finely and chop the onion into small pieces.
3. Put them in a pan with the vinegar, salt, and spices. Peel and chop the apples, add them to the pan, and bring them to the boil. Reduce the heat, cover, and simmer it for half an hour.
4. Remove the plum stones, chop them, and add them to the pan. Stir in the sugar and cook for about 40 minutes, occasionally stirring.
5. Spoon the mixture into the jars, cover with a fabric or plastic top, and tie it on.
6. Store for a month before you use it to let the flavors develop.

Spiced Apple Chutney

Ingredients:
- 1 ½ cups white sugar
- 1 ½ cups white vinegar
- 4 tart green apples
- ¼ cup golden raisins
- ¼ cup dried apricots, diced
- ¼ cup shallots, diced
- 5 thick fresh ginger slices
- ¼ tsp. Aleppo or red pepper flakes
- 1 star anise
- 2 minced garlic cloves
- 1 tsp. kosher salt
- ½ tsp. yellow mustard seed

Instructions:
1. Put the sugar and vinegar in a large pan and whisk. Peel, core, and chop the apples into small, even chunks. Add them to the pan with the ginger, raisins, apricots, shallots, star anise, and pepper flakes. Stir and heat to a simmer.

2. Reduce the heat to medium-low, add the salt, mustard seeds, and garlic, and simmer for 40 to 45 minutes. The apples should be soft, and the liquid should be reduced.
3. Remove the pan from the heat and let it cool before lifting the star anise and ginger out.
4. Place the chutney in the refrigerator until chilled and season with pepper flakes and salt.

Chapter 7: Canning and Preserving Plant-Based Food

It sometimes seems like all or nothing when you grow your own fruit and vegetables. Sometimes, you get little for your effort, and at other times, you'll have so much you won't know what to do with it. That's where safe food preservation comes in, allowing you to grow enough throughout the growing season to feed yourself all year.

Canning

Canning is a popular method of preservation.
https://unsplash.com/photos/a-wooden-shelf-filled-with-lots-of-jars-of-food-bhni1zsPiio

One of the most popular methods of preservation is canning, and there are two main types:
- Water bath
- Pressure

Which one you use is determined by the food's acidity levels – low-acid foods must be water bath canned, while high-acid foods must be pressure canned.

Food Acidity
Low-Acid

These include most vegetables (not pickled foods or acidified tomatoes) and meat. They do not have sufficient acid to stop bacteria that would be killed by boiling water. The worst bacterium is Clostridium botulinum, which is responsible for a deadly food poisoning called botulism. Using a pressure canner destroys the botulinum spores because it uses much higher temperatures.

High-Acid

These include most fruits, although tomatoes are on the low-high acid borderline and must be processed using vinegar, citric acid, or lemon juice to increase their acidity levels. It also includes pickled and fermented foods, jellies, and jams (unless made with low-acid vegetables.)

Let's look at the two popular canning types.

Boiling Water Bath Canner

The food is heated when the jar is submerged in boiling water. A temperature of 212°F must be maintained to kill bacteria and enzymes, and each recipe will specify the exact processing time. These are usually given for altitudes of 1,000 feet or lower above sea level. Remember that water requires a lower temperature to boil at high altitudes, so processing time is usually longer.

Pressure Canner

Heat is applied when the canner is sealed, forcing pressure to build up inside. This creates steam, which pushes air out, and when the canner vents are shut, the canner only contains pressurized steam. This is much hotter than boiling water, but again, adjustments for altitude are required.

Canning Procedures

No matter which type of canner you use, there are set procedures you must follow.

- **Choose Your Jars and Lids:** although commercial single-use jars may also be used, Mason jars are ideal. However, you may have trouble getting lids to fit these. You can get jars in many sizes, but most recipes use quart and pint jars. You can use larger half-gallon jars for juices, and most jars can be reused if cleaned and sterilized. Lids should be flat metal disks with a separate screw band and edged by a sealing compound. Never reuse lids. Bands can be used again if they are not rusty or damaged.
- **Hot or Raw Pack?** Raw packing requires the food to be packed raw into the jars. However, as it can float inside the jar, it may discolor within a couple of months. Hot packing requires the food to be boiled, simmered, and filled into the jars hot. This ensures the air is removed from the food, shrinking it and stopping it from floating. It also allows more food to go into the jars.

General Guidelines

- Always use well-tested recipes, i.e., those from the USDA.
- Mason jars can withstand higher temperatures than single-use jars, especially for pressure canning.
- Jars should never be filled cold. Always heat them in simmering water or a dishwasher.
- Always leave the right headspace. Jams, juices, and jellies require ¼-inch, pickles, tomatoes, and fruits require ½-inch, and veggies and meats require an inch. Too little means food may be forced into the seal, stopping it from working, and too much can weaken the seal with a low vacuum.
- Use a bubble remover or another plastic tool to remove air bubbles.
- Clean the jar rims before putting the lids and bands on.
- Two-piece lids should be used – a screw band and a flat disk.
- Lids should only be finger-tight.
- Use a jar lifter to get the jars into and out of the canner.

- Do not tilt jars when you move them.
- Use the correct processing method per the recipe.
- Adjust pressure and time for altitude.
- All jars should be set a couple of inches apart on a thick towel or heat-proof board to cool.
- Never turn the jars upside down or retighten the screw bands.

Water Bath Canning Procedure:
1. Follow the guidelines.
2. Half-fill the canner with water and preheat it: 140°F for raw-packing and 180°F for hot-packing
3. Put the jars on the rack inside the canner.
4. Pour in more water if needed to ensure the jars are submerged by at least an inch.
5. Put the lid on and leave it on during the process.
6. Turn the heat right up until the water is boiling. Reduce the heat to a gentle boil.
7. If the water stops boiling, turn the heat back up and start over.
8. Once the time is up, turn off the heat, remove the lid, and leave the jars for 5 minutes. Then, lift them out and place them somewhere to cool.

Pressure Canning Procedure:
1. Follow the guidelines.
2. Put 2 to 3 inches of water in the canner and put the jars on a rack at the bottom.
3. Tighten the lid and heat the water to boiling, venting the steam for 10 minutes. Then you can close the petcock or add weights.
4. Add the pressure or weight regulator.
5. The pressure must then rise and be maintained for the time stated in the recipe. If the pressure drops, you must start over.
6. Once processing is finished, turn off the heat and let the pressure cool until the pressure is at zero. Wait for 2 minutes, and then remove the regulator. Wait a further 10 minutes, and then remove the lid and set the jars aside to cool.

Testing the Seal

Jars must cool for 12 to 24 hours, and then you should press on the lid center. If it moves, the jar is not sealed. If it doesn't, and you cannot remove the lid once the band is removed, the jar is properly sealed.

Storage

- Take the screw bands off and wash the jars.
- Store somewhere dark, cool, and dry at a temperature of 50 to 70°F.

Dehydration

Dehydrated foods are becoming more popular.
https://unsplash.com/photos/flatlay-photography-of-citrus-and-dragon-atzWFItRHy8

Dehydrated foods are becoming very popular and are not just good for snacks. You can serve them with dips, add them to stews and soups, or even use them as pizza toppers. It's a simple way to preserve certain foods and make your own fruit leather, vegetable or fruit chips, and jerky.

1. **Pick Your Produce at Its Peak:** Contrary to popular belief, you cannot use old, bruised, or damaged food. Drying food concentrates the flavor, so always start with food at its prime.
2. **Clean and Peel:** Wash fruit and vegetables thoroughly and peel them. You don't have to do this, but peel does become tough during dehydration.

3. **Slice Thinly:** Use a sharp knife or mandolin to slice the food into uniform pieces, ¼ to 1/8-inch thick. That way, they all dehydrate at the same rate.
4. **Dip It in Citrus Water:** This is really only for fruit that goes brown, such as apples and bananas. Dip it in a solution of 50% water and 50% lemon juice for about 10 minutes. Then remove and pat dry.
5. **Blanch It:** This is for starchy foods like potatoes and peas. Boil for a few minutes, and immediately put them in an ice bath to stop the cooking process.

Using a Dehydrator:

These are simple to use. Simply layer the food on the trays, set the dial, and turn it on. Fruit should typically be dehydrated at 135°F, while vegetables require 125°F. Times will vary according to the food being dehydrated, its initial ripeness, slice size, and the day's humidity.

Using an Oven:

While this is an option, be aware that ovens may be hotter and don't always heat consistently. The oven must be preheated to its lowest temperature, and cooling racks must be placed on lined baking trays. Layer the food on the trays and cook. You must monitor them carefully as drying time is shorter than a dehydrator. You will also need to rotate the baking pans for even drying.

Storing Dried Produce:

Cool the food completely and store it in airtight containers, like Mason jars, snap-top lids, or Ziploc bags. They should be stored in a dark, cool place. Shake a container after a few days, and if you see any moisture, place the food back in the dehydrator or over for a while. Once dried properly, food can last for several months.

Pickling and Fermenting

Fermenting and pickling are age-old preservation methods commonly but incorrectly discussed interchangeably. While there is some overlap between them, they are two separate methods.

Pickling:

Pickling is defined as preserving food in acidic brine and has been used for thousands of years as a preservation method. It doesn't take much to do, either, making food taste sour, with a softer texture.

You can quick-pickle foods easily by boiling a solution of water, vinegar, sugar, and salt and pouring it over the prepared vegetables. You can also add a pickling spice mix. This method destroys microorganisms and bacteria.

Fermenting:

This is also a simple method and is older than pickling. It uses yeast, bacteria, or other microorganisms to turn food carbs into alcohol. There are two fermentation types:

- **Alcoholic:** commonly used in beer, wine, and bread. Yeast and bacteria break pyruvate down into carbon dioxide and ethanol.
- **Lactic Acid:** the pyruvate molecules break down even further into lactic acid.

Fermented foods are rich in probiotics, making them an excellent choice for the digestive system. Kimchi, miso paste, and sauerkraut are popular fermented foods.

The Pickling Liquid:

- **Vinegar:** use white or white wine vinegar for a clear color, but red wine, rice, apple cider, or a combination of vinegar also works well.
- **Water:** needed to dilute the vinegar.
- **Salt:** a preservation agent that adds flavor.
- **Sugar:** helps balance the acidic taste of vinegar.
- **Whole Spices:** use coriander seeds, mustard seeds, dill seeds, fennel seeds, celery seeds, cumin seeds, star anise, allspice, clove, and/or black peppercorns.
- **Fresh herbs:** dill works best, but basil and cilantro also work.
- **Garlic cloves:** you can also use onion slices.

How to Pickle Veggies:

1. **Wash Your Vegetables.** Peel them if needed and slice those that need slicing.
2. **Pack the Vegetables into the Jars.** Layer them with herbs, whole spices, and raw onion, ginger, or garlic.
3. **Heat the Pickling Liquid.** Do this while filling your jars, allowing the liquid to boil.

4. **Fill the Jars.** Pour the liquid over the vegetables, leaving ½-inch headspace, wipe the jars, and put the lids on.
5. **Cool Them.** Leave the jars on the counter to cool for a few hours, then refrigerate for 12 to 24 hours. This will allow the flavors to develop. If you can leave them for a few days, it's even better!

Sauerkraut

Ingredients:
- 4.4 lb. firm white or pale green cabbage, outer leaves removed
- 6 tbsp. flaky sea salt OR 3 tbsp. coarse crystal sea salt
- 1 tsp. whole peppercorns
- 1 tsp. caraway seeds

Instructions:
1. Sterilize a large bowl with boiling water, then wash your hands thoroughly.
2. Thinly shred the cabbage and layer it in the bowl with the salt. Massage the cabbage gently for five minutes to get the salt into it, then stop for five minutes before massaging again. The cabbage should end up reduced and coated in brine.
3. Stir in the peppercorns and caraway seeds and cover the cabbage surface with plastic wrap, pressing the air bubbles out. You should see the brine level rise to cover the cabbage.
4. Cover the entire bowl with a clean towel and leave it somewhere dark and cool for 5 days or more. The longer you can leave it, the better the flavor, so try to leave it for 2 to 6 weeks or until the bubbles have stopped.
5. Check it daily to allow gases to be released and stir it to get rid of the bubbles. Remove any scum that forms.
6. Make sure you keep the temperature consistently cool, 64 to 68°F. Any cooler and fermentation will take longer – and any warmer, it may turn moldy.
7. When the cabbage is how you want it, pack it into small sterilized jars and store it in the refrigerator for up to six months.

Dill Pickles
Ingredients:
- 8 to 10 pickling cucumbers
- 4 cloves of garlic, halved.
- 2 tsp. mustard seeds
- 2 tsp. peppercorns
- 2 cups water
- 2 cups distilled white vinegar (5%)
- ¼ cup cane sugar
- 2 tbsp. sea salt
- Fresh dill sprigs

Instructions:
1. Slice the pickles into 4 pieces lengthwise to make spears, or slice them thinly widthwise to make pickle chips.
2. Divide them between 4 8-oz. or 2 16-oz. sterilized jars.
3. Divide the mustard seeds, garlic, and peppercorns between the jars and add dill to each one.
4. Combine the sugar, salt, water, and vinegar in a pan and heat over medium heat. Stir to dissolve the salt and sugar, and then leave it to cool a little.
5. Pour the liquid into the jars, put the lids on, and leave to cool for a few hours.
6. Refrigerate for 5 to 6 days for spears or a couple of days for chips.
7. These can be stored in the refrigerator for several weeks.

Root Cellaring

In the days of old, people didn't have refrigerators to store their food. Indeed, many didn't have the luxury of electricity, so they used root cellars to store their food, especially root vegetables. Root cellars use the earth's natural insulation and cooling properties to stop food from going off in the summer and freezing in winter.

Root cellars aren't just for root vegetables, though. You can also store your canned foods, pickles, fruit, other vegetables, dairy and meats, homemade juices, and alcoholic drinks.

The beauty of a root cellar is that if the electricity goes out, you don't have to worry about food spoiling. You can harvest from your garden all summer long and store it away for winter, saving you a ton of money in electric and shopping costs.

Basic Requirements

You need to consider a few things if you are thinking about having a root cellar on your homestead:

- **Legal Requirements:** You must consult your local building department to see the requirements for building a root cellar. Some areas will require you to have a building permit.
- **Temperature:** Your climate is important because root cellars use the earth for cooling and insulation. If you live in southern states where the climate is typically hotter, root cellars don't work so well. The temperature in the cellar must be kept at between 32 to 40°F.
- **Location:** Be careful about where you build your cellar. For example, don't plan to build near the sewage or septic system, where the water table is high, or in areas where flooding is a problem. Also, consider the size of the cellar. On average, they are typically 8 x 8 feet.
- **Humidity:** Root cellars need 85 to 95% humidity, no lower. If your cellar floor is gravel or dirt, this won't be a problem because the earth will do it for you.
- **Ventilation:** Root cellars need air circulation, so your cellar must be airtight with the right ventilation system.
- **Darkness:** Your root cellar needs to be dark, or your veggies may start sprouting, change color, lose their nutrients, or even go off. If your cellar has windows, they must be covered in blackout material, and the lights should only be turned on when absolutely needed.

Different Types of Root Cellar

There are several types of root cellar, some attached to your home, others not:

Basement

If you have a basement, you're in luck. Part of it could be turned into a root cellar, but there are a couple of recommendations to follow:

- The northeast corner foundation walls should be used as two sides of the cellar.
- The other two walls should be built using boards and studs
- The ceiling, interior walls, ducts, pipes, and door must be insulated to ensure the room stays cool

Hole in the Ground

This is your traditional cellar and is simple enough to build. Dig a large hole in the ground or, if you live on hilly land, into the sides of a hill. Reinforce the sides so you have a room and put a doorway in. Obviously, it's a bit more complex than that, but you get the idea.

Barrel

If space is limited, you could have a simple barrel cellar. This just requires burying a large garbage can or barrel into a hole in the ground. The food should be stored inside the can with straw, earth heaped around the outside, and the lid covered with mulch or straw and a plastic sheet. This will keep root vegetables fresh no matter how cold it gets.

Storing Your Vegetables

Always suit your produce before you store it. You should only store mature, firm vegetables and fruit that are undamaged and unblemished. Damaged, rotten, or overripe food can contaminate everything else, so these should be composted or used immediately.

Use baskets, mesh bags, or other containers with good air circulation, and make sure there are no sharp edges. Some foods should not be stored together or even near each other. The following summary will give you an idea of winter veggie storage requirements:

- **Cabbages:** best kept in the ground, covered in heavy mulch. Alternatively, hang them from their roots in an external building. It must not be stored indoors as it gives off ethylene gas. Store at 32 to 40°F, 80 to 90% humidity.
- **Garlic and Onions:** cure or dry them somewhere ventilated for 2 to 3 weeks. Then, hang them in the cellar as they are or in mesh bags. Keep the cellar at 40 to 50°F, 60 to 70% humidity.
- **Potatoes:** should be cured for a few weeks first. Layer firm, undamaged, or unblemished potatoes in one layer somewhere dark. Cover them with newspaper and leave them for 2 weeks at 45 to 60°F, then store them in total darkness in wooden boxes

(covered), paper bags, dry sand, or tin buckets. If they begin sprouting, they are too warm. Do not store near apples or onions, as these hasten sprouting.

- **Squash and Pumpkins:** Leave the stems on when you harvest, as this lessens the risk of disease. Store them in a cool, dry place, 40 to 60°F, 60 to 70% humidity, in mesh bags or placed on shelves, not touching each other.
- **Root Crops:** can stay in the ground covered in mulch and withstand temperatures of 28°F or above. To store in a cellar, leave an inch of stem and top and layer in boxes or baskets. Put sand or sphagnum moss between each layer and store at 32 to 40°F, 80 to 90% humidity.
- **Citrus:** store at 40 to 50°F, 60 to 70% humidity in mesh bags or baskets. Oranges can go as low as 32 to 34°F.
- **Pome Fruits:** this includes pears and apples, and they need to be stored in very cold temperatures, as near to 32°F as possible, at 80 to 90% humidity. Store in boxes layered with moss or sand, in plastic bags with holes, or individually wrapped in paper and layered in boxes. As they emit ethylene gas, they cannot be stored near any other crop.

Section Three:
Living Off the Land II: Animals, Meat, and Dairy

Chapter 8: The Buzz on Beekeeping

The world is abuzz with talk of the bees dying off, but why is this such a hot issue? The answer is simple – bees are one of the most important pollinators for food crops. However, because so many chemicals are used on commercially grown food, and partly due to climate change, the bees are dying. If people don't repopulate them quickly, food crops will disappear, too.

Grocery stores would stock fewer vegetables and fruit. You wouldn't be able to buy anything with honey, coffee, chocolate, almonds, etc. There would be little beef and milk because the plants used to feed the cattle would not be pollinated. They even help trees and plants grow, providing the planet with oxygen. It's safe to say that, without bees, we would struggle to survive. And, if no plants were pollinated, they wouldn't reproduce, and other wildlife would lose a food supply and shelter.

How to Keep Bees

To keep bees, you need certain equipment.
https://unsplash.com/photos/person-in-white-pants-and-white-shoes-standing-on-brown-wooden-table-LZ3Ffa_Yl7E

Keeping bees is not something you can just decide to do. You need specific equipment before getting the bees, and it's tough for a beginner to know where to start.

The essential equipment you need:
- **Hives:** obviously, you need a hive, and there are two primary ones to choose from. Langstroth hives are the most common wooden sections stacked on top of one another and containing hanging frames. Top bar hives contain frames hanging from the top bar, with honeycomb hanging from the frames.
- **Bee Suit:** these are designed to cover you completely to protect you from bee stings. A complete suit is made from heavy fabric and includes a hat, veil, gloves, and boots.
- **Hive Tool:** this is a small tool but one of the most important, as it can help you separate the boxes, get the frames out safely, and get the excess propolis and wax from the rest of the hive.
- **Smoker:** an essential piece, a smoker keeps the bees calm as the smoke covers their pheromones. This lessens the risk of being stung by upset bees.

- **Bee Brush:** this helps you get the bees off the frames when needed, get them into boxes when they swarm, or even get them off you.
- **Queen Catcher:** if you need to keep the queen separate for a while, this tool helps you do it without hurting her.
- **Feeder:** bees are good at feeding themselves through the spring and summer months, but they need food throughout fall and winter, too. A feeder allows you to feed them on dry sugar, honey, or sugar syrup.
- **Capping scratcher:** this narrow-bristled wire brush helps you uncap the wax cells when getting the honey out or checking for disease.

As a beginner, you are best buying ready-made hives. Later, you can buy kits if you want to expand.

Placing the Hive

This depends on property size, how near your neighbors are, access, landscaping, etc. Bees are pretty flexible, so choose somewhere that lessens the risk of disturbance, etc., your neighbors, you, and your livestock:

- Face the hive entrance south or southeast. Your hive needs at least 6 to 8 hours of sun from early morning.
- To provide the bees with a clean way to and from the hive, ensure that there's a clear path of about a 10- to 15-foot area in front of the hive.
- It's essential to create a fence around the hive to keep any predators at bay.
- Find an ideal location where there won't be any disturbances.
- Use cinder blocks or a proper hive stand so the hive is 12 to 18 inches off the ground. This keeps skunks and other animals out of the hive.
- Weight the hive lid down with a large rock to stop the wind from blowing it off or animals from getting into it.

Get Your Bees

Beekeeping firms, local beekeeping associations, and farm supply organizations all offer online ordering for bees. You will have to inquire about ordering and delivery schedules from nearby companies. Orders are typically placed in January and delivered in March or April.

Ask local keepers if they can recommend reputable sources, and when you find a supplier, ask whether the bees have been treated with antibiotics. You don't want to purchase those because the bees may have American foulbrood disease, which can wipe out your colony.

The most common bees are Carniolan, Russian, and Italian, each with advantages and disadvantages.

- **Carniolan:** these are quiet, gentle bees who spend their winters with low numbers of bees, building up very quickly in the spring. However, they are prone to swarming if they don't have enough room to expand. They are less likely to suffer disease and are excellent at keeping their hives clean.
- **Italian:** You'll find that Italian bees are less defensive than other species and are the most widely preserved and dominant bee race, particularly in North America. They don't abuse propolis, the resinous glue that bees use in their hives, they don't swarm frequently, and they weather the winters well. They do over-brood, too, and they go through honey fast.
- **Russian:** Russian and Italian bees are similar, but Russian bees tolerate varroa mites better. They wait until pollen is available before they raise their broods and stop when the pollen begins to disappear. They also keep a queen cell active all year.

Bees can be bought in two different forms: as a package (bee box) or as a nuc (nucleus colony). The latter are about the size of a shoe box, with a single queen kept in a separate queen box, and are typically constructed of wood, plastic, or screening. They often arrive with enough sugar water to last them for a few days, but they can't stay in the box longer than four or six days after they were placed inside. Nucs offer a significant benefit over bee boxes since they often contain frames with honey and wax, a few bees and their offspring, and the queen.

Be aware that both types can come with diseases, and there will usually be varroa mites. Carefully inspect the bees for anything odd before you introduce them to your hives.

Install the Bees

1. Choose a sunny, warm day, above 55°F, with no wind if possible.
2. Give your bees a 1:1 sugar-water solution. Remove a frame from the hive and install an inline feeder if you have one. Place your

Boardman feeder at the hive's entrance if you have one.
3. From the feeder, you can transport three or four frames.
4. Get dressed in your bee suit, hat, veil, gloves, and boots.
5. Using a mix of water and sugar, spray the box. This will simultaneously feed the bees and help them calm down.
6. Open the bee box and remove the sugar can, leaving the queen undisturbed.
7. Take out the queen box, clear the worker bees from it, and set it above the frames. To keep the other bees where they are, replace the sugar can.
8. Insert the queen into the colony. Ascertain that she is not facing the box's corked end, then carefully take the cork out. Put a tiny marshmallow into the hole as soon as possible. The queen will be released for the following several days as the remaining bees gnaw into it. By doing this, they ensure that they perceive the queen as their queen by imprinting with her pheromones.
9. Using an elastic band or stapler, suspend the queen box between the two middle frames. The screen of the queen box should not be covered in any way. To feed the queen, the bees must have access to her via the screen.
10. After the queen is inside, shake the bees into the hive and take the sugar can out of the bee box. To get as many out as you can, do this firmly.
11. Replace the frames slowly, making sure all the bees are out of the way so they don't get injured or killed.
12. To ensure that any bees that are left leave and head towards the queen, you can place the bee box inside the entrance to the hive.
13. Now, place the inner and top covers on.
14. Weight the top down, find the smallest opening and install the hive reducer.

You have to spend as much time as possible outside the hive over the following few days. The bees can turn on the queen and obtain a new one if they are agitated excessively during the settling process. Seeing dead bees emerge from the front entrance is one of the best indicators that your hive is settling. This indicates that the worker bees have made their home and are tidying the hive.

Open the hive a few days after installation and ensure the queen has been let out. If yes, the queen box can come out. If not, quickly open the box and get it back into the hive. Replace the inner and top covers to stop her flying away.

Within 7 to 10 days, you can look at the frames and see if the queen is laying eggs.

- You should check on the hive every 10 to 21 days.
- Make sure your beekeeping equipment is kept clean to stop diseases and pests from spreading.
- Place a shallow birdbath or dish for water near the hive and keep it filled. Make sure the bees cannot drown in it.
- Fill your garden with flowers and flowering plants so the bees can collect the pollen and nectar. This helps them grow and reproduce healthily, so plant plenty of flowers that bloom at different heights and times throughout the season.
- Keep a close eye on your hive and monitor for diseases and pests that can decimate the hive.

Bee Products

Most people assume that bees only make honey, but, in actual fact, they make many other products, some of them incredibly valuable. Here's what you could get from your beehives:

Honey

Honey is quite complex and is made when honey bees collect sweet deposits and nectar from trees and plants, modifying it and storing it in honeycombs in the hive. This is a great food source for your bees, but you can also gather it yourself, as long as you only take what you need and leave enough in the hive for the bees to feed on.

Nectar

Nectar has a high level of sucrose and comes from nectaries, which are plant glands. It also has a high moisture level, and the bees use that moisture in their honey production process. Nectar is one of the most common sources from which bees get energy.

Beeswax

Beeswax is one of the purest forms of wax and is all-natural. Bees must visit more than 30 million flowers and consume 8 to 10 lb. of honey to make just one pound of beeswax. To do this, the young bees in the colony get together in a large cluster to make their body temperatures rise. Beneath their abdomens are wax-producing glands, and these begin to secrete wax slivers no bigger than a pinhead. The worker bees take these wax slivers and transport them to where they are needed in the hive to form the honeycomb. Each bee will produce around 8 slivers of wax in 12 hours, so you can see the colony's level of patience in building their comb to keep their young safe and fed.

Pollen

Only worker bees have a "pollen basket" in which they gather pollen from flowers and take it to the hive. The pollen is an excellent protein source for bees to raise a healthy brood. Sometimes, excess pollen may be harvested and used as a health supplement. It can be consumed in small amounts, such as sprinkled on a yogurt. If you need to hand pollinate flowers, you can also use the excess pollen, but be aware that it has to be used within a few hours of being collected, or its potency decreases rapidly.

Bee Bread

Bees make this from honey and plant pollen formed into granules and stored within the honeycomb. Bee bread is commonly consumed to stop allergies from pollen, and Olympic athletes also consume it to boost their immune system, recover quickly after a training session, and boost their performance. It isn't recommended that you try to get this from your hive, though. Leave it for the bees to feed on.

Propolis

The word is Greek in origin and translates loosely to "defending the city." Otherwise known as bee glue, bees make it from tree balsams, saps, and resins, and it is often used to seal up any cracks appearing in the hive. If wild bees swarm into a tree hollow, they may even use the propolis to make a way into the hive. Dwarf honey bees coat the branch where their nest is in propolis to stop ants from getting to it, too. Propolis has medicinal qualities and is often taken as a health supplement in capsule form. You might even find it in some toothpaste and cosmetics.

Royal Jelly

This is secreted by honey bees and used to feed their larvae and the queen. Worker bees are the only ones with the hypopharynx glands needed to secrete the jelly. When the workers decide a new queen is needed, usually because the old one has died or is weak, they pick a number of small larvae and place them in cells constructed especially for the purpose. They then feed them on a lot of royal jelly, triggering the larvae into developing "queen morphology," including the ovaries they need to lay the eggs.

All larvae are fed on the jelly, no matter whether they are destined to be workers, drones, or queens. This only happens for three days unless they have been selected as queens. You can harvest royal jelly from your hives if your hive has several queen cells, but only when the queen larvae have reached 4 days old. This is because the queen cells are stocked full of it, whereas the other larvae are fed at certain times. If your hive is managed correctly, you could theoretically collect half a kilo over 5 to 6 months. However, it perishes quickly and must be immediately stored in a freezer or refrigerator until needed or ready to sell.

Bee Venom

Worker bees inject venom when they sting in a bid to protect their colony and themselves. The venom is called apitoxin and is a clear, colorless liquid with proteins that can cause a mild or severe allergic reaction. Bee venom has long been used in natural healing, but collecting it is neither easy nor advisable for those with no experience.

How to Harvest the Honey

The whole point of keeping bees is pollinating your plants and producing honey. The first honey harvest for a new beekeeper is exciting and is actually quite simple, so long as you have the right equipment.

The Right Equipment

Before you start, you need the right equipment, so make sure everything is there first.

The essential tools you will need to harvest a few hives are:
- A hot knife or special uncapping knife to remove the wax caps from the cells.
- A fine strainer to catch debris and dirt in the honey.
- Food-safe buckets to store the honey in before transferring it to jars.
- A tray for the wax cappings – a proper capping tray or a sterilized baking tray.
- Food-safe honey jars with tight lids – plastic or glass will do.

- Measuring jug to transfer the honey to the jars – it must have a spout.

Find Somewhere Suitable

When you have everything together, find somewhere to extract the honey. You can do it outside, but indoors is better as you won't be plagued by bees, wasps, and other insects. Ensure your chosen space is clean and big enough.

Using a Honey Extractor:

1. Make sure your honey is ready to harvest. This means checking that every hive frame is full and the honey is smothered in a white wax cap.
2. Lift the frames and lay them out on a stable surface – make sure it is clean first. Don't take all the honey, as your bees need to feed over the winter.
3. If you have a honey extractor, set it up in the center of your space. Place the mesh strainer on the edge of your bucket and situate it beneath the extractor's spout.
4. Put the capping tray on a bench, plug the knife in, and leave it somewhere stable to heat up.
5. The frames must be uncapped before you can use the extractor to get the honey. Work out how many frames your extractor can hold and remove the wax caps on that many frames. Use the knife to scrape the wax from either side of the frame – do it over a tray, as this can get messy.
6. Fit the frames into the extractor and turn it on or, if it is manual, turn the handle. The extractor tank will spin, and the centrifugal force sucks the honey out, and it drips out through the spout.
7. After a while, check one side of the frame. If it is empty, turn the frame around and repeat to empty the other side.
8. Remove the empty frames, uncap and add more, and repeat until all the frames are empty.
9. It can take an hour or so for all the honey to drain, depending on how many frames you are processing and the heat in the area.
10. During this time, you can sterilize your jars and lids. When the honey is collected, take the collection bucket away, replacing it with another smaller one for any drips.

11. Remove the strainer from the bucket, set it aside, and use your jug to collect the honey and pour it into jars. Tighten the lids firmly, label each jar with the date, and store the honey in a dark, cool place.

Once you have extracted all the honey, you will have a valuable pile of wax that you can use to make other products. However, you will need to remove any traces of honey from the wax. If you have a small amount, you can do this by placing the wax in water and heating it. The wax will float on top of the water, and the honey will stay at the bottom. Once it has melted, let it cool completely and separate the wax from the water. Store it in a clean container with a lid until you are ready to use it. Do NOT use a good pan for this, as beeswax tends to stick, and never use the same pan for anything else. Hopefully, you will be doing this regularly throughout the season, so have a specific pan you use only for this.

Cleaning up is pretty easy, even though honey extraction is messy. However, as honey dissolves in water, simply wash everything with hot, soapy water and clean cloths.

Extracting by Hand:

If you don't have access to an extractor, you can crush and drain the honeycomb to extract the honey. It is an effective method, but it does take time.

1. Remove the frames and break the honeycombs into small chunks. Place them into a food-safe, clean bucket.
2. Using a large masher, crush the comb down, reducing it to a quarter or third of its original volume.
3. Place a large colander over another bucket and pour the crushed honeycomb into it. It will take about an hour to stain through.
4. Next, pour the strained honey through a fine-meshed sieve into another container – you can also line the sieve with cheesecloth if you want. The honey that comes through will be clean and can be transferred straight to the jars and stored.
5. You will be left with the wax, but this can be processed in hot water, as detailed above.

Learn how to make products, such as candles, lip balm, soap, or polish, or use it to waterproof leather, such as horse harnesses, to protect them.

How to Protect Your Bees from Pests and Diseases

Careful monitoring is the best way to stop diseases and pests in your hive. However, that means knowing what to look for and how to deal with them. These are common ones that attack honeybees:

- **Aethina tumida**

Better known as small hive beetles, these are black-brown and tiny, just ½ a centimeter long. They can travel up to 20 km and quickly infect a hive, causing serious damage to colonies. The beetles are clever, imitating the behavior of the bees so they can enter the hive unharmed. Once in, they lay eggs, which hatch within six days and feed on the pollen, beeswax, and honey, destroying the comb.

Look For:
- Tiny beetles in the comb or hiding in cool parts of the hive
- A rotten-orange smell or slimy combs
- Small eggs in cracks and corners
- Honey fermenting or dripping from the comb cells
- Larvae eating the brood and food stores
- Clumps of larvae in the comb cells or frame corners

Prevention:

You cannot stop these beetles from getting into your hives. All you can do is find them quickly and interrupt their life cycle before they go too far. They are a notifiable pest, so immediately let your local government bee safety and pest management department know.

- **Chalkbrood**

Caused by a spore-producing fungus called Ascosphaera apis, the bees ingest the spores while they eat, which ferment in the larvae's guts, starving them. Foraging bees usually bring the disease back to the hive, and infection spreads very quickly. Left untreated, it can weaken a hive, allowing other diseases and pests to take hold.

Look For:
- Bee larvae coated in a chalky, cotton-like substance
- Black or grey larvae
- Mummified larvae on the pollen traps or the entrance to the hive

- Brood nests filled with dead larvae

Prevention:

Keep your colony healthy and happy so they can fight off the infection. If the infection has gone too far, you may need to replace the comb frames or re-queen the hive with a much stronger brood.

- **Nosemosis**

Sometimes called Nosema, this disease is caused by Nosema apis and Nosema ceranae, both single-cell fungi. It is the most common disease in full-grown bees and is incredibly contagious. It can lead to a much shorter lifespan. The spores are ingested through food, cleaning, water spots, and other places in the hive. They go into the stomach, damage the epithelium, and affect the bee's ability to digest.

Look For:
- Worker bees with swollen bellies in the hive
- Fewer bees in the hive or a struggle to survive the winter
- Lack of honey and lower brood populations
- Dysentery

Prevention:

A healthy hive is the best way to prevent this disease, so ensure your bees have the best nutrition. Rotate the hives every few years to keep them strong, and don't move hives when you don't need to keep stress levels low. It is a notifiable disease and must be reported to your region's Department of Agriculture.

- **Wasps**

The last thing you want is wasps crowding your hives, as they can kill your bees.

Look For: Large groups of wasps

Prevention:

Make the area around the hives unattractive to wasps by growing mint, eucalyptus, citronella, and wormwood. Also, make sure there is no meat or sweet food waste around the hives, as these attract the wasps. Pick up fruit as soon as it falls off the trees, and keep your compost bins and rubbish bins tightly lidded. You can also hang wasp repellants, such as Waspinators, near the hives.

- **Wax Moths**

Two types of wax moth can attack your hives: Galleria mellonella (greater) and Achroia grisella (lesser.) Both eat larval remains, pollen, and beeswax and will lay their eggs in gaps or cracks. Left alone, they will take over the whole hive.

Look For:
- A broken-down comb.
- Cocoons that look like white webbing.
- Bald parts in the brood.
- Dark feces, shaped like cylinders, on the hive floor.

Prevention:

Again, a healthy, clean hive is the best prevention because strong bees have a better chance of kicking the moths out. Make sure your hive has no gaps, the roof is tight, and that you don't add another entrance, as these are all ways the moths can get in.

- **American Foulbrood**

AFB is caused by Paenibacillus larvae, a bacterium that forms spores. It can kill bees and entire colonies, even the strongest ones. Spores are spread from hive to hive by infected equipment, and it is incurable. The spores can stay active for at least 50 years.

Look For:
- Patch, irregular patterns in the broods
- Dark, greasy, sunken caps on broods
- Dead larvae – usually a liquid mass
- Decomposing larvae

Prevention:

Follow best practices and keep your equipment clean. Inspect the combs twice yearly, and if an outbreak occurs, the infected hives or colonies must be destroyed, along with infected equipment.

Obviously, the best way to prevent disease and pests is to keep your hives and equipment clean. Build a robust and healthy colony, inspect the hives regularly, and contact pest control if you see anything suspicious.

Ethical and Sustainable Beekeeping

While these are separate concepts, they are interrelated. In beekeeping terms, ethics is about ensuring your bees' welfare and that of their surroundings. Sustainable beekeeping is about keeping the ecosystem healthy and understanding that every action has a consequence.

Ethical Practices:

Ethical beekeeping covers quite a bit of ground, but, in simple terms, it's more about what you don't do rather than what you do. It's about keeping your impact on the bees' lives to a minimum, along with your impact on their natural surroundings. It's also about keeping the bees as stress-free and danger-free as possible while allowing them to go about their lives naturally.

Ethical beekeepers:

- Do not take more from a hive than it can spare
- Only harvest at certain times of the year
- Don't have huge hives, so the bees don't have too much to do
- Don't use chemicals or synthetic replacements for honey
- Keep swarming to a minimum to stop introduced bee species and native fauna from competing for resources

Sustainable Practices:

Sustainable beekeeping is about the relationship between humans, bees, and the entire ecosystem, and beekeepers work to protect the whole bee population, not just their own. Being aware of the environment is important as beekeepers must understand the bees' needs and the dangers they may face.

Sustainable beekeepers:

- Understand the need bees have for a diverse array of flowering plants and will plant different kinds of trees and plants to feed the bees at different times of the year.
- Avoid the use of chemicals, especially pesticides, in their garden, as these can harm the bees.
- Try to keep native rather than introduced species, as the latter can spread disease and cause damage to the ecosystem.
- Adapt their hives to cope with climate change. Bees cannot cope with rapid changes in temperatures, so they ensure the hives are

properly adapted to help them.

Once your hives are set up, keeping bees is relatively simple, and it is an incredibly rewarding part of your homestead. They don't just provide you with honey and beeswax. They also pollinate your crops, ensuring your harvest is a bountiful one.

Chapter 9: Livestock Selection and Care

There will come a time when you'll want to start keeping livestock on your homestead, especially if you want to become more self-sufficient. Most livestock, such as cows, sheep, pigs, ducks, chickens, etc., produce food, and some can even be used to work your land for you.

However, before considering what animals to have, you must prepare your land for them.

Preparing for Livestock

Raising livestock requires a lot of preparation.
https://unsplash.com/photos/brown-goat-on-blue-metal-cage-DdbQGZEkZM

This is the biggest job you will have to do:
- **Shelter and Fencing:** Every animal has its own fencing and shelter requirements, so do your homework first. For example, you can keep chickens in a coop with housing and ventilation, while pigs must be in a secured area, as they are good escape artists. Larger animals, such as cows, need an acre of land per animal.
- **Water:** All animals need water, and large animals can go through several gallons daily. Having a natural, sustainable source of water on your homestead is ideal. Watering your animals in the winter is harder, especially in cold winters where everything freezes solid. To assist with this, you can use electrically heated water buckets and bowls or equip larger troughs with floating heaters. Additionally, be ready to stock up in advance if you use an electric pump to draw water from a well because you might not be able to reach it during storms and power outages.
- **Managing Manure:** keeping livestock means manure and a lot of it, depending on the livestock you keep. However, it can all be used on the homestead. Rabbit manure is a great garden fertilizer, while chicken manure should be added to the compost heap or left to age for a while before using it in the garden. Cattle and horse manure must be left to rot down before using it. Ensure you have an area set aside for manure management on your homestead.
- **Predator Protection:** this is a big problem, especially if your homestead is isolated. In order to protect your cattle and keep the predators away, make sure your fencing and shelters are strong enough to fend off threats from a number of predators, such as wild dogs, hawks, coyotes, and many more.

Choosing Your Livestock

Deciding what livestock you want is also important. There's no point in keeping animals just for the sake of it because that could be an expensive mistake. Here are some of the animals you could consider keeping:
- **Chickens:** The most common livestock or homesteaders, chickens produce eggs, meat, and manure, and you can use them to clear patches of land of insects and weeds.

- **Ducks:** One of the easiest fowl to keep, ducks don't need much space and produce meat, eggs, and manure. They need a constant water source and a secure pen as they are slow creatures that predators can easily get. This includes hawks, so make sure the pen has a covered area.
- **Quail:** These are small birds that are good for meat and eggs. One of the cheaper birds to keep, quails need little space, but be aware that the eggs are small, and you won't get much meat – you might need a lot of them. They are also temperamental layers and, as they are good flyers, they won't be suitable for a free-range homestead.
- **Pheasant:** Larger birds that produce meat and eggs but not in the same quantities as ducks and chickens. They are harder to tame, though, so they may be more challenging, and their laying is irregular.
- **Turkeys:** Larger birds that supply meat and eggs but do need a fair amount of space, especially if you intend to free-range them. They also need a water source and secure pen and shelter.
- **Geese:** Many homesteaders keep geese for their meat and eggs, but also because they make excellent guard animals. They can be vicious, though, and you should have warning signs up if you intend to let them roam your property.
- **Rabbits:** These are an excellent source of meat and breed well, so you can have a steady supply. You should have one male and several females, but be aware they will fight, so be prepared to keep them separate.
- **Goats:** Good for beginners, goats are relatively easy to keep and an excellent source of milk and meat. Do your homework on what breeds to keep, as some are better than others. They are cheap to buy but need a good, secure shelter as they can cause destruction. They also need a regular food and water supply and are escape artists.
- **Pigs:** Another escape artist animal, pigs are quite easy to keep, so long as you keep them in secure pens. Because of their quick reproductive cycle, a piglet bought in the spring is prepared for butchering in the fall. All they really need is a little area, some cover, and water, and they'll eat almost anything.

- **Cows:** If you have plenty of space, cows are not a bad option, but you need secure fencing and pastures to rotate the animals around. Keep in mind you need an acre per animal, and they need a never-ending water supply. If you intend to keep them through winter, they need a lot of care but are an excellent source of milk and meat.
- **Horses:** These are not typical homestead animals because they don't produce food. However, they are good workers and can be used to plow your land, carry goods, and move stuff around your homestead, and they produce plenty of manure. However, they are not cheap to keep and require daily grooming, plenty of water, and a good shelter, not to mention strong fences.

Non-Traditional Livestock

Once your homestead is established, you may want to consider keeping more exotic animals. However, you must be aware that, as these are not local to your area, they are more expensive to purchase and need a lot of care. Some of the commonly kept are:

- **Ostrich:** Excellent source of eggs, meat, and feathers, but they need a lot of space and can be dangerous.
- **Emu:** This animal is smaller in size than the ostrich and provides a decent amount of eggs, feathers, and meat.
- **Llama:** The llama is famously known for its coat of fur. It is used for a number of things, such as carpeting, fabric, and even rope.
- **Alpaca:** The alpaca's fur is also useful as it can be utilized for soft fleece and clothes.
- **Elk:** Large animals that need a lot of space but are an excellent source of meat and fur. However, they are quite dangerous.
- **Bison:** Raised for their meat and fur and are easier to look after than cattle.

Before You Buy

Some animals may look cute, but they are not always easy to care for, not to mention the expense of purchasing and caring for them. Consider a few things before you decide to go down this route:

- Are you eager to have an animal that can be useful for a number of things or even as just a pet?

- Do you think you have the heart to butcher your animals even after raising them?
- Have you ever raised animals for meat before?

Answering these questions will help you determine whether you are ready to keep productive livestock and which breeds to go for. It will also help you determine the type of shelter and fencing you need.

There are some other important questions you need to consider:

- **Can I Keep Animals in My Area?** Some areas don't allow livestock, while others will have rules on what you can and can't keep. Check your local restrictions and requirements before you purchase livestock.
- **Have I Got the Space?** Some animals need more space to roam and live. Consider bees, chickens, and rabbits if you only have a small piece of land. If you have more space, you can consider larger animals.
- **Am I Away from Home a lot?** They may be there for food, but livestock depends on you for everything. If you need to travel, ensure you have someone reliable and knowledgeable who can look after things in your absence.
- **Are They Productive?** You need animals that produce food or other products for a self-sufficient homestead. If they don't, they should be useful in other ways, such as a dog guarding the animals, horses for plowing and as transport, etc. When you look after the animal, you should get something in return.
- **Can I Eat Them?** Not everybody relishes the thought of eating animals they've raised, much less doing the butchering themselves. If that's you, choose animals you can raise for milk and eggs instead. However, you should consider meat animals and, if you can't do the butchering yourself, learn how or find someone who can do it for you.
- **Does the Climate Suit Them?** Most animals will adapt to their surroundings and climate, but not all are suited to certain environments. It's imperative to pick breeds that thrive in the climate you live in.

Legal Considerations

You just need to obtain general liability insurance if your homestead is operated exclusively for you and your family. An EIN (Employee

Identification Number), which is required for anyone selling goods or services, is necessary if you plan to sell any things.

However, the authorities would view your homestead as a commercial company if it is set up to sell products like honey, eggs, and vegetables. This implies that you need to incorporate your company. Additionally, you'll require appropriate homesteading contracts and a website.

If you have livestock, things get a little trickier. You are liable if your animals escape and cause an accident or damage any other property besides your own. You have a legal requirement to provide secure fencing so your livestock cannot escape and also to protect other people from being harmed by your livestock. This doesn't just mean having good fences and locked gates. It also means having plenty of clear signs around the property to warn people of the dangers. You should also consider proper liability insurance in case anyone makes a claim against you for damage or injury.

Building a Shelter

Various livestock shelters exist, and what you need depends on your animals. Before you build, consider the environment and where you want to build it. Then, consider whether the animal will be safe and comfortable.

Ensure your buildings are strong enough to withstand inclement weather like snow, strong winds, or heavy rain. Ensure it has a sloped roof so that snow and rain can run off, and have a water collection or diversion system so the ground doesn't get excessively boggy. Lastly, ensure it has plenty of ventilation.

Three common types of shelter are:
- **Barns:** one of the more popular shelters, barns work well for large animals. They are typically constructed of metal or wood, with a pitched roof, providing space, stalls, tack rooms, feeding areas, and anything else your animals need.
- **Run-in Sheds:** these work well for small herds of sheep or cows and are constructed with wooden walls and corrugated roofs.
- **Tilt Shelters:** these work best in areas that get strong winds. They have pitched roofs and are designed to bend without breaking.

Build or Buy?

If you choose to build, you need time, resources, and help. It isn't easy to build a shelter if you don't know what to do. New homesteaders may find it easier to purchase prefab shelters. However, before you build or buy, ensure your zoning laws allow them.

Materials

If you build from scratch, you will need certain materials. Flooring should be made from concrete as this is easier to keep clean. The main skeleton can be constructed from concrete blocks or solid wooden posts, so long as they go at least a foot into the ground to provide stability and stop animals from knocking them over.

Roofing can be made from tin or corrugated metal, and the sides can be constructed from wooden planks, wire fencing, cattle planks, or even bricks if you have the budget. Consider what is best for your animals and climate, but ensure you don't leave any sharp edges or nails sticking out.

Different Designs

You can choose from several designs, depending on your livestock and budget. Most people start with a run-in shed or open-air barn. However, while these can provide shelter, you must consider that your livestock needs protection from the elements and a clean, dry place to feed. If your climate is hot and dry, these shelters can work. However, they are not ideal if you get strong winds, rain, and cold winters. Even in hot climates, you need to consider insect and pest populations.

Enclosed shelters are the better option, but you must ensure adequate ventilation and shade from the sun and rain.

Building Tips:

- Make sure external doors are tightly shut at night to keep predators out.
- For cold nights, you need a heat source to stop your animals from getting too cold or getting ill.
- Ensure adequate ventilation. Windows on opposite walls and roofline openings are good ideas.
- Poultry should have a covered outdoor area to get fresh air and warmth but where they are protected from the sun.
- Ensure your building materials have no sharp edges and that your roofline gaps don't allow other animals to get in and attack your

livestock. You can stop this by placing sheets of strong wire fencing over them.

Supplementing with Feed

It doesn't matter how many acres of pasture you have or how much extra garden veg you have to feed your animals. You still need to supplement with animal feed. Here's a quick guide for the most common homestead livestock:

Chickens:

Chickens don't just produce meat and eggs. They also give you plenty of manure that contains nitrogen and are a great pest control system. They also eat weeds and can till the soil for you.

They have specific nutritional requirements:
- Plenty of fresh, clean water
- Fats
- Proteins
- Carbohydrates
- Minerals
- Vitamins

They are omnivorous, so they eat vegetation and animals. They need room to forage for insects, seeds, and grass, but they do not have teeth, which means they need grit in their diet to help break their food down. They also need specific chicken feed containing the right mix of nutrition for their age.
- **Age 0 to 6 Weeks:** starter feed
- **Age 6 to 14 Weeks:** grower feed
- **Age 14 to 20 Weeks:** developer feed

Layers need a specific layer feed containing a high calcium level to provide strong shells for their eggs. Purchase a feed with a minimum of 2.5 to 3.5% calcium.

Breeders need a specific layer/breeder feed to get optimum nutrition levels.

Geese:

Geese are great lawnmowers and can clean up areas of your land quicker than you can, especially under bushes and trees. They also make good guardians, keep predators away from other livestock, and provide eggs.

Geese are mostly happy to eat whatever they can find, especially the grass, and will also eat cracked corn and wheat grain, vegetables, fruit, and insects. They have serrated edges on their beaks, making them good grazers, but it will hurt if they bite!

For their first 4 weeks, feed goslings on a good starter food with at least 20% protein. Then, they can move to grazing, supplemented with a 16% protein feed.

Cows:

Cows provide milk and meat and are great grazers, keeping long grass down. They produce plenty of good manure to make fertilizer from, rebalancing your soil's carbon levels.

Most of their diet is fresh grass, but they will also eat cut and dried grass (hay) when needed. They also eat corn, oats, and wheat stems and leaves. Beef cows require fats, proteins, carbs, minerals, vitamins, and plenty of water, while dairy cows need food with a dairy meal in it to boost milk production. Dairy cows also need up to 50 gallons of water per day.

Pigs:

Pigs are great for meat production but also work as rototillers on your land. Let them at your land, and they'll till it in no time, digging up rocks, eating roots, and clearing the land of grubs and other insects.

They are omnivorous and will eat many different foods, but you should ensure that 20 to 40% of their food intake is pasture and grain. They will also eat vegetables, fruit, mushrooms, grain, nuts, seeds, insects, and eggs.

They need access to fresh, clean water 24/7 and protein, carbs, minerals, fats, and minerals, all found in an organic, balanced pig feed.

Top Tips for Animal Husbandry

There's more to raising animals than providing food, water, and shelter. For successful animal husbandry, there are certain best practices you should follow:

1. There's No Place for Impulsiveness

This is the first rule of animal husbandry to learn and understand. Never be impulsive in buying or obtaining livestock when you don't have the land or means to care for it properly. Do your homework and take your time. Work out why you want the animals. Do you want chicks because they look cute or because you want the meat and eggs they produce? Are lambs just fluffy, cuddly bundles, or do you visualize the wool and meat they provide? Piglets might be cute, but have you got the space for a full-grown animal or two? What about goats? If you want them for milk and meat, don't buy the first one you see. Research the best breeds for your environment and for the purpose you want them for.

Do you have the money and time needed for proper husbandry? How much money and time are needed for each breed you want to get?

Do you have sufficient space? Can you build a proper shelter and still have enough space for grazing and roaming?

Which types of livestock can you raise together?

Are you prepared to hand-rear young animals if something happens to their mother?

Put simply, know exactly what you want and why, and ensure you can give it the best care possible.

2. Secure Housing

Your livestock must be safe and secure in their new home, giving them somewhere to escape the weather, keep away from predators, and roam about. All animals have specific needs, but you must be realistic. You cannot keep the animals in cramped conditions, as they need to be able to move about freely. Too many animals in too small a space can also lead to illness and disease.

They will need the right kind of shelter, too. Chickens need a coop, goats and sheep need a weatherproofed shed or barn, while cattle and horses need a large barn with plenty of space.

3. Check Your Fences and Check Again

Shelters and space are just one part. Your animals also need secure fencing to keep them in and people and other animals out. What you get depends on your livestock. Chickens will get away with wire fencing as long as they have a secure building to roost in at night. Other animals may need electric fences. Once your fence is up, you must check it daily and immediately make necessary repairs.

4. Water Access

All animals need water, but you don't want to be out there every 5 minutes with buckets. Ensure your animals have constant access to water, especially larger animals who would benefit from a natural water source on the land. You must also be prepared for cold winters, have water stocked up ahead of time, and be ready to thaw water if it freezes solid.

5. The Best Nutrition

Healthy, well-cared-for animals provide healthy food for you, so their nutrition is one of your top priorities. Whether you purchase it or make your own, you must ensure your livestock gets the right nutritional needs and supplement feed where needed.

6. Get a Good Veterinarian

Ensure you do this before you have any livestock because they can help you choose the right ones and give you plenty of advice about their care. You also need to know you can call your vet any time, day or night, should an emergency arise.

7. Think about the Environment

You must consider whether your chosen livestock can handle environmental factors. Some breeds are happy outside, while others aren't. Some animals handle the heat, while others are better suited to colder temperatures and can happily survive a harsh winter with the right care. You also need to consider that you will be looking after them, so ask yourself if you can handle extremes of temperature every single day. This isn't the type of job where you can decide not to go to work because of the weather!

8. Inspect Your Livestock Daily

Successful husbandry takes time and effort. That means checking on all your animals every day. You'll need to understand what issues each breed may face so that you can quickly carry out your inspections. That way, you stand a better chance of picking up on a disease, illness, or injury and dealing with it immediately.

9. Have a Good First Aid Kit

A first aid kit for animal husbandry should contain the following at a minimum:

- A syringe
- Antibacterial/antifungal spray
- Bandages
- Corn starch
- Cotton swabs
- Electrolyte powder

- Gauze
- Hydrogen peroxide
- Neosporin
- Saline solution
- Scissors
- Towels
- Tweezers
- Vaseline

You also need to check with your veterinarian to see if there are specific medications, creams, and ointments you should keep on hand.

10. Be Ready for the Poop

There will be a lot of it, and you must keep your livestock areas clean. Have a plan in place for cleaning it up regularly (at least twice daily) and what you will do with it. Some choose a composting system where the poop can be turned into rich fertilizer. Others heap it up and let it age before spreading it on their land.

No matter what you do, husbandry can only be successful if you keep these tips in mind and be prepared to put in the time and effort. If you are not 100% committed, don't even consider keeping animals.

Sheep Shearing Techniques

If you have sheep and want to harvest their wool, you need to learn how to shear them. You may have some idea how to do it, but if you don't do it properly, you can stress the animal and potentially cause injury to you or it.

Here are some shearing tips to help you:

- **Be Confident:** This is one of the most important things to learn about shearing. Confidence is everything, so never fumble with the shears, and don't second-guess yourself, or you are more likely to go wrong. You should also get into a good rhythm to move easily between the sections that need shearing.
- **Don't Make Second Cuts:** If you don't perform the first cut properly, you are forced to make a second cut. This isn't ideal as these bits can't go onto the rolled fleece, as they weaken the yarn and cause pilling. Experience is the only thing that will teach you how to avoid making second cuts.

- **Keep the Skin Taut with Your Non-Dominant Hand:** This allows the shears to move easily, and you'll cut closer because the skin is stretched. This is important when shearing fine-wooled sheep, as they have wrinkled skin, and it's not so easy to see through the wool.

Know Your Sheep

This helps you shear your sheep better:

- Is she fat or thin? This helps you move the shears at the correct angle for her body shape.
- How many teats? Some have two, while others have four, and you need to know so you can steer clear of them when shearing.
- Is the sheep healthy? If yes, there'll be plenty of lanolin in the wool that melts and makes it easier for the shears to slide through. If not, the lanolin is thick, which makes it difficult to move the shears.

Milking Techniques and Tips

Knowing how to milk your cattle properly is essential to getting a good supply of milk and ensuring your animals are comfortable. Good techniques are also essential to the animal's health.

How to Get Sufficient Milk:

How do you know you are getting as much milk as your cow can give? The answer lies in the animal's health, comfort, and calmness.

- **Health:** you will nearly always get more milk from a healthy cow than an unhealthy one. Check your herd for signs of illness, stress, or disease, and if anything looks wrong, contact your vet. Also, make sure your animals receive their vaccinations.
- **Comfort:** cows need good food and plenty of fresh water, shelter from the elements, and space to roam and graze. If a cow is cold, she shivers, using energy to stay warm. If she is too hot, she won't eat so much. Both result in a lower milk yield.
- **Calmness:** never shout at your cows or hurt them, or they get distressed and scared. That means less milk, so be calm and gentle with them, and you'll get more milk.

Keep Things Clean

You must keep the enclosure, shelter, and milking area clean. If they are dirty and wet, it can lead to disease. The cows must also be kept clean, especially just before you milk them. Brush them down to remove dirt and dust, use warm water to wash their teats and udders gently, and dry them with paper towels.

Cut long hair away from the teats and ensure your fingernails are short and clean. Wash your hands thoroughly before milking, and do not cough or spit anywhere near the milk. Also, ensure you have no open wounds or sores on your hands.

Keep It Regular

Cows get used to a routine, so try to milk them at the same times every day. Even a half-hour deviation can result in less milk and cream.

Check for Mastitis

Place a piece of black nylon stocking over a bucket. The first spurt of milk from each teat should go through this. If the milk is watery or lumpy, the cow likely has mastitis, and you cannot use it. Mark the cow and get her treated immediately. From then on, milk her last.

Milking Technique

There's more to milking than grabbing a teat and squeezing it! Following the right technique ensures your cows are comfortable and you get the optimum amount of milk.

1. Lubricate your hands and hold the top of the teat with your thumb and forefinger. Slide them down the teat gently, but do not pull on it. This technique is only for the first milk spurt and is called "stripping."

2. Now, use your whole hand to enclose the teat, keeping the forefinger and thumb around the top. Starting at the top, squeeze each finger on the teat individually, then release the forefinger and thumb and repeat. This is called "expressing" and mimics how a calf feeds from the cow. If your cow is not in calf, take all the milk, or the cow will be uncomfortable.

3. Repeat the first step to get the last spurt when you have all the milk from the udder. Do not tug the teat, as this can cause mastitis, which isn't comfortable for the cow. You should also never use your fingers and thumbs to squeeze out the milk, as this is uncomfortable, damages the top bit of the teat, and can cut milk

production.

4. When you have finished milking and transferred the milk to a suitable container, wash the milk can and bucket with cold water, followed by boiling water with sodium carbonate dissolved in it - use ½ a cup for every 10 liters of water. Drip-dry them upside down, wash all the cloths in soapy water, and rinse and let them dry.

5 Beginner's Butchering Tips

It won't be easy to do if you have never butchered an animal before. Ensure you learn how to do it properly if you don't intend to let someone else do it.

1. Use the Right Tools

At the very least, you need a *honing steel* to help you keep your knife blades sharp while butchering. It's a given that you will hit the bone occasionally, and keeping a steel nearby allows you to keep the blade sharp.

Next, you need a meat hook. This allows you a good grip when you are separating the meat. You will also need a boning knife, preferably with a well-made steel blade that is semi-flexible.

Next is a machete to cut through big bits and a hand saw to get through the bone. A larger hand saw is also ideal to help you slice larger cuts.

You will also need a tough glove - consider chain mail - to save your hands from the inevitable nicks and cuts.

2. Decide on Your Cuts

Are you after steak? Roasts? Ribs? Or maybe you want hamburgers or stewing meat. Choose the cuts you want and research how to make those cuts. The best place to go is a butcher, as they can tell you or even give you a visual guide on each cut. Knowing how to make your cuts will give you a good idea of how to process the meat. Once you have the cuts you want, you can use the rest of the carcass to make exploratory cuts and learn from your mistakes. You can turn scraps into stewing or ground meat, so there won't be any waste. However, try not to turn the loin meat into scrap. This is the bit on the top of the spine and is classed as high-value meat.

3. Learn to Be Organized

Make sure you have enough space to process the meat. Several buckets and a way of labeling each cut is also recommended. You can tack bits of paper to them if you like. Once you've processed enough animals, you'll learn what each cut looks like, thus speeding up the process.

4. No Two People Process the Same Way

That means that no one process is better than any other. When you have an entire carcass, you can afford to make a mistake or two, and you'll learn your own process along the way. So long as you know where the cuts come from, you'll find a comfortable way of processing that makes sense to you.

5. Control the Carcass Temperature

This can make all the difference in how you process the meat. It will be much harder to cut and take forever if it is too cold, even partly frozen. If your meat is frozen before you process it, let it thaw for 24 hours in a warmer room – not too warm, though. Some meat, especially beef, is left to hang for at least 18 to 21 days to ensure it is tender enough. If the meat is warmer, it will be easier and faster to process, and peeling the muscles back and finding the joints is also easier. Never try to process a frozen carcass, as it will take too much time and won't do your hands any favors, either.

Five Tips for Raising Livestock for Food

You may not make much profit when you raise animals for food, but it gives you the highest-quality meat and can save you money in the long run. You only need enough land per animal, good, healthy feed, plenty of water, and time to care for the animal.

As a homesteader, you will probably consider livestock at some point in your journey, especially if you want to be self-sufficient, and at least you will know what went into your food and how the animal was treated. When you purchase meat from a store, it may meet the USDA standards, but you may be shocked to learn how the animals were kept and raised before being slaughtered.

Raising your own animals means knowing they were treated kindly, humanely, and looked after well. You know what it was fed and whether it was healthy. Many homesteads raise cows for milk and beef, raise pigs, egg-laying birds, and meat birds, and there's good reason for doing all of this.

Grocery store prices are rising fast, and most of the meat, eggs, and milk you can buy there comes from animals raised in less-than-ideal conditions. Listen to the news, and most days, you will hear about an outbreak of one disease or another, affecting prices and food quality, so it's little wonder that more people are raising their own food supplies.

Make no mistake, though. Raising your own livestock for food is not easy. It takes time, money, effort, and a lot of blood, sweat, and tears, but the rewards are huge. Humans have been doing this for centuries, and, in many places, it is a way of life. Homesteading is about going some way toward making it a way of life for you.

Here are some tips to help you make the best of it.

Choose Your Livestock with Care

You have several options if you decide to raise animals for meat:

- **Meat Chickens:** Ready to butcher at eight weeks old.
- **Laying Hens:** Some breeds can be used for eggs and meat. If you buy hens, you have an instant source of meat.
- **Beef Cattle:** You need an acre per cow for grass-fed, but you'll get enough meat from one cow to last an entire year.
- **Pigs:** Must be fenced in, but you'll get a decent amount of meat if you put in the effort for 6 months. They will eat any food scraps from your garden, but you will likely have to purchase a supplemental feed.
- **Sheep:** Easy to raise, so long as they have plenty of pasture land to graze on, and you'll get a decent amount of meat. Be aware that mutton takes more cooking than lamb.
- **Goats:** Goats don't need to graze on high-quality pasture land and are easy to keep. They produce a decent amount of meat, but it must be cooked long and slowly, or it will be quite tough. Kid meat is more tender, but you don't get so much of it.
- **Rabbits:** Ready for butchering from 8 weeks old, and as they breed so well, you get a plentiful supply.

Think about the Endgame

If you can't decide what animals to raise, think about what you want. Knowing your endgame will tell you what animals you need. If you want a small dairy farm, start with goats or cows. You'll need a beef cow if you only want or have space for one animal but want a lot of meat. If you want a quick way of getting meat on the table, you'll want rabbits and meat chickens, as both can be butchered at eight weeks.

Consider the Expense

You know that raising livestock isn't free, nor will it be cheap. You have the initial layout of purchasing the animals, building shelters, fencing, and then there's feed, which is an ongoing purchase. You may be forgiven for thinking it must be cheaper just to go buy your meat from the store.

Raising some animals will be cheaper if you have plenty of grassland, especially cattle, but you need enough for them to graze from spring through to fall. You will have to purchase haylage (fermented grass) to supplement during the winter, but this is much cheaper than bales of grass. However, if you don't have enough land, you need to consider the cost of buying food.

Young chicks and piglets need a heat lamp in the colder days, but cattle, goats, and sheep tend to be easier. So long as they have shelter, they can keep out of the wind and rain and are reasonably self-sufficient.

You will need to consider the cost of repairs to your fences and shelters, another ongoing cost. In terms of breeding for succession, see if you can come to a deal with a neighbor who may have a bull, a billy goat, a ram, or other male animals that you can breed yours with.

Weighing up all the costs, work out the cost per pound of raising your own against purchasing from a store. Also, consider that your own animals are likely to be raised organically, whereas you cannot guarantee that with store-bought meat.

If you find feed and butchering too expensive, see if you can go halves with a neighbor. Food bought in bulk is generally cheaper, too. If you live in a homestead or farming area, you'll find plenty of people willing to help out and share costs.

Consider the Time

Raising livestock is not a 5-minute job. You need to be committed big time because your animals will need daily care. Even if they graze during the summer, you must provide fresh water regularly. You could consider an automatic stock tank valve. This ensures your water tanks stay full without you needing to fill them. Large animals drink an awful lot of water in a day, and it's a tough job if you have to keep trenching out to the field to top it up.

You also need to inspect your animals every day for signs of stress, disease, illness, and injury. You must keep an eye on your fences and ensure any damage is repaired straight away, and this has to be done regularly.

If you have to go away, you need to hire someone who is capable and knowledgeable to care for your livestock, and regardless of the weather, whether you are ill or if anything comes up, you still need to get out there and tend to those animals.

Think about the Emotional Investment

Livestock are not pets. Yes, you will look after them and get to know every one of them, but you must keep telling yourself that they are food, not pets. You must learn to distance yourself emotionally while still providing a clean, safe environment that's as close to their natural habitat as possible. That means raising them organically, grass-fed, and pasture-raised.

You will benefit from knowing that your animals are happy and healthy. Stressed animals translate to tough meat, while happy animals are tastier and better for you.

Raising livestock is well worth the effort, provided you have the time and money to do it.

Chapter 10: Making Your Own Dairy and Meat-Based Food

It doesn't matter whether your homestead is a tenth of an acre or a hundred because you can be self-sufficient in many ways. By definition, homesteading is all about self-sufficiency and sustainability. It's about growing your own food, raising livestock for food, preserving your food to keep it for longer, and may even involve making your own clothes. It means living within your means and doing your bit to fight climate change.

Growing veggies can be done small-scale in containers and raised beds or large-scale with huge fields of food, and you can preserve it in many ways to keep your family fed the whole year. In the same way, you can raise enough livestock to feed your family without relying on grocery stores for meat, dairy, fish, etc., that you know is not organic and has often been raised in poor conditions. When you raise animals for eggs, meat, and dairy, you have everything you need to survive on your journey to self-sufficiency.

Humans started life as hunter-gatherers, fishing and hunting for food to sustain them. Being a homesteader means you are getting back to basics, learning how to provide your family with good, nutritious, organic food in a sustainable lifestyle that helps you, your community, and the climate.

Let's look at how to process and preserve some of the meat and dairy you have raised on your homestead.

Handling and Processing Dairy

How to Handle Raw Milk

There has been much scaremongering about raw milk, but it is incredibly safe and clean. It only becomes unsafe when it comes into contact with contaminated containers, surfaces, and tools.

When handling raw milk, you must ensure you sterilize everything it may come into contact with, including your hands!

Storing Raw Milk

Your storage system must be in place before you get your dairy animals. Each cow can produce up to 35 liters of milk daily, and if you have nowhere to store it, you'll waste it or spend a ton of time turning it into something else – every day!

Raw milk will stay fresh for up to two weeks when properly refrigerated.
https://unsplash.com/photos/mason-jar-filled-with-smoothie-S1HuosAnX-Y

Milk will stay fresh and sweet for up to 2 weeks when properly refrigerated. After that, you can still culture or use it in baking, but it won't taste so nice when you drink it.

Using Your Milk

No matter what dairy product you see at the store, there's a pretty good chance you can make it at home. Here are some simple recipes to get you started.

Queso Blanco Cheese

This is one of the easiest cheeses to make and has a lovely, mild flavor.

Ingredients:
- 1 gallon of milk.
- ¼ cup apple cider vinegar.

Instructions:
1. Heat the milk to 185 to 190°F, stirring to stop it from burning.
2. Remove it from the heat when it gets to temperature, and stir in the apple cider vinegar a bit at a time, stirring until the curds and whey have separated.
3. Put the curds in a cheesecloth and hang them for a couple of hours until no more liquid drips out and the cheese has solidified and cooled.
4. Slice, crumble, or cube, and enjoy within a few days.

Yogurt

Ingredients:
- 1 small tub of plain, active yogurt.
- A quart of milk

Instructions:
1. Heat the milk until bubbling. You should see skin forming on the top.
2. Pour it into a quart glass jar and leave it to cool down to about 115°F.
3. Stir in 2 to 3 tbsp. of plain yogurt and stand the jar in a pan. Add hot water from the tap (as hot as possible) until it reaches the yogurt level.
4. Leave it to set for 8 to 12 hours, and then cool it in the fridge.

In the future, use a little of this as your starter in the next batch instead of buying it.

Sour Cream
Ingredients:
- 1-quart milk
- 1 pack sour cream starter

Instructions:
1. Heat the milk gently to 86°F. If it's a hot day, pour it into a jar and set it in the sun until it warms up.
2. Add the starter and shake or stir it.
3. Leave it at room temperature to thicken up to 24 hours.
4. Label it, and refrigerate.

Butter
Ingredients:
- 2 cups heavy cream off the milk.
- ½ tsp. salt

Instructions:
1. Use an electric mixer to mix the cream and salt on speed 8 or 9. Mix for 10 to 15 minutes until the bowl is coated in solids.
2. Scoop the solids into a ball and hold it under cold running water, squeezing it until the water runs clear.
3. Simply store it in the refrigerator and use it within a few weeks.

The liquid left in the mixing ball is buttermilk, which you can use in baking.

Handling and Processing Meat

When you handle raw meat, you must wash your hands thoroughly before and after to avoid cross-contamination, and never use the same equipment you used on meat with any other food.

There are a couple of ways to process meat, the most popular being curing.

Curing Meat

You can do this in two ways. The first method is salt curing, and the second is brining, typically used for food smoking. For example, you would need 2 to 3 tbsp. of salt to cure a 5 lb. steak, although sometimes sugar is used too, but that depends on the recipe.

Brining can also preserve, ferment, and pickle meat, enhancing the flavor, color, and texture. Highly salted water is usually used in brining. If you have a 15 lb. joint of meat to brine, submerge it in just enough water to cover it and stir in 6 tbsp. of salt. As soon as the meat has soaked in the brine all through, it can be smoked until tender.

Brining and curing are the secrets to perfectly smoked meat, imbuing it with a smoky flavor that keeps it tender and moist. The trick is not to put too much meat in the container to ensure plenty of room to absorb the brine or salt completely.

If you don't have access to a smoker, there are a couple of other ways to use up some of your meat apart from freezing it.

Homemade Sausage

Ingredients:
- 3 ½ lb. lean meat – whatever you have available
- 1 ½ lb. fatty meat – pork belly, shoulder, etc.
- 2 tbsp. kosher sea salt
- 1 tbsp. black pepper
- ½ cup ice water
- Sausage casings

You will also need a meat grinder and a sausage stuffer

Instructions:
1. Chop all the meat and fat into small chunks – they must fit the grinder.
2. Put them in a bowl, mix with the salt, and refrigerate overnight.
3. The next day, freeze your blade, auger, and dies for an hour and soak 15 feet of sausage casings in warm water.
4. Mix the meat chunks with the black pepper and grind them using a coarse die. If the meat is still below 38°F, you can grind it again using a fine die. If not, freeze it for 20 minutes and grind it again.
5. Freeze the meat – it must be as close to freezing as possible. Remove it from the freezer, add the ice water, and work it with your hands for 60 to 90 seconds until you have a sticky ball.
6. At this stage, you can use this as sausage meat or make sausage links. If you want links, put the meatball into the sausage stuffer, slip on the casing, and leave a 4-inch tail.

7. Remove the air by cranking the stuffer down and "milking" the casing out of the tail. Then, fill the casing with the meat. Repeat until you have used all the meat – don't make them too tight.
8. Make the links. Press one end and tie it off, then pinch a link off about 6 inches in. Twist it away from you a couple of times to seal it, then repeat, twisting it towards you this time. Do this down each coil and tie the last one off.
9. Check for air bubbles in the links, using a needle to prick them out, compressing the meat gently to fill in any gaps. Hang the sausages for 24 hours before consuming, freezing, or storing.

Chicken Liver Pate
Ingredients:
- 1 lb. chicken liver
- 1 cup milk of your choice
- ¼ cup unsalted butter
- 1 tbsp. olive oil
- 3 to 4 minced garlic cloves
- 1 cup finely chopped onion
- 2 whole bay leaves
- ½ tsp. dried thyme
- ½ tsp. salt
- ½ tsp. pepper
- ¼ cup brandy

Instructions:
1. Put the liver in a bowl, cover it with milk, and stir gently. Refrigerate for at least 2 hours.
2. Melt the butter and olive oil over medium heat, and cook the onions until soft. Add the minced garlic and cook for a minute or so, but do not allow the onions and garlic to go brown.
3. Add the liver, salt, pepper, and bay leaves and cook until the liver is just a little pink inside and browned all over.
4. Remove from the heat, add the brandy, and place it back on the heat. Set the brandy on fire and cook until the liquid has gone. Remove the bay leaves and let it cool.

5. Once cool, blend to a smooth pate and refrigerate for several hours before eating.
6. This pate also freezes well.

Chapter 11: Preserving Meat, Dairy, and Eggs

Making delicious dishes with meat and dairy is one way to preserve it. However, if you don't want to use it immediately, there are other ways to preserve it for long-term use.

Preserving Meat

Meat can be safely stored for years if it is properly preserved.
https://unsplash.com/photos/sliced-meat-beside-silver-knife-Xcdxbjx7MFg

You can safely store meat for years if you preserve it properly. The obvious way is to freeze it, but there are some other methods you can use, some thousands of years old.

Freezing:
Before freezing meat, it must be prepared to prevent freezer burn and prolong its life.

- Wrap meat and poultry in a couple of layers of plastic wrap, then a layer of aluminum foil and a plastic bag designed for freezer use.
- You can also use a home vacuum sealer to shrink-wrap your meat.
- Airtight plastic or freezer-safe glass containers are also safe to use.
- Do remove as much bone as possible, as these are a contributing factor in freezer burn, and the more bone you store, the less room you have for meat.
- If you want to freeze sliced meat or patties, place a sheet of parchment paper between them to stop them from sticking together.

Label Your Packages
Each package or container must be labeled with the type of meat, date, and whether it is raw or cooked. Try to group items, i.e., all the chicken together, and separate cooked from raw.

Use the meat in date order, i.e., earliest first. That way, you won't waste food that has freezer burn or is out of date.

Know How Long to Store It
Contrary to popular belief, meat cannot be stored in a freezer forever:

- **Uncooked Meat:** chops, steaks, etc. – 4 to 12 months
- **Uncooked Ground Meat:** 3 to 4 months
- **Cooked Meat:** 2 to 3 months
- **Lunch meat, Ham, and Hotdogs:** 1 to 2 months
- **Poultry:** cooked and uncooked can be stored for 3 to 12 months
- **Wild Game:** 8 to 12 months

Ensure your freezer temperature stays at 0°F or lower.

Thaw Your Meat Properly

It is rarely a good idea to cook your meat from frozen. Knowing how to thaw it properly can save you from food-borne illnesses.

- **In the Refrigerator:** Things like full turkeys can take up to 24 hours to thaw, so plan ahead. This is the safest method.
- **In Cold Water:** Ensure the meat is in airtight packaging and submerge it in cold water. The water must be changed every half-hour until the meat is thawed.
- **In the Microwave:** The meat must be cooked immediately afterward, as microwaves are notorious for uneven thawing and can even begin to cook some of the meat.

Check your meat before you cook it. If it smells off or has gone an odd color, discard it.

Salt Preservation:

As mentioned earlier, salt can be used to preserve meat, which is one of the oldest ways.

- Use proper curing salt – you can get this online.
- Meat should be put into airtight bags or containers and covered completely in salt. The best way is to start with a layer of salt, add meat, another salt layer, meat, and so on.
- Containers should be stored at 36 to 40°F for a set time – do not allow them to freeze.
- The formula for how long to cure your meat is seven days for each inch thick. For example, a ham weighing 12 to 14 lb. and 6 inches thick should be cured for 42 days.
- You can keep salt-cured meat for up to four months without refrigeration if you store it in airtight containers or bags.
- Always rinse off excess salt before you cook it.

Dehydration:

Making jerky is an excellent way of making protein-rich snacks and needs only an oven and stovetop or a dehydrator.

- Slice the meat into thin strips with a 1 cm. x 1 cm. cross-section.
- Boil it for 3 to 5 minutes – this removes all bacteria.
- Drain the meat until it has dried.

- Bake for 8 to 12 hours on the lowest setting in your oven, or use a food dehydrator (follow the manufacturer's instructions.)
- When dried properly, the meat will feel leathery, hard, or sticky.

You can store jerky for up to two months in an airtight container without refrigeration.

Smoking:
Smoking is a great way of preserving meat, and it adds flavor.
- Meat should be salt-cured first, as this will allow for longer storage.
- Smoke in a proper smoker for 7 hours at 145°F or 5 hours at 155°F. Do not go over 155°F as the meat will cook rather than be smoked or dried.
- Be aware that some cuts, such as brisket, may take longer, up to a day in some cases.
- A meat thermometer should be used to check the internal temperature before taking the meat from the smoker. Poultry should be 165°F, roasts, steaks, and chops 145°F, and pork and ground meat 160°F.
- Flavor the meat using mesquite, hickory, or cherry wood chips.
- Store for up to 3 months in an airtight container.

Canning:
Canning is another popular method of preserving meat, but the process must be followed exactly for safe preservation.

Always Use the Right Tools:
- Meat should be processed using a pressure canner as the steam sterilizes the meat, cooking it and sealing it in the jars.
- Always use Mason or other good canning jars with no chips or cracks.
- Never open the canner until the processing is complete and it has cooled and naturally depressurized. Do not force it by running cold water over the canner.
- You can store canned meat for up to a year in a cool, dry place.

Canning Poultry/Rabbit:
- **Hot Pack:** Steam, boil, or bake until the meat is two-thirds cooked. Pack it into the jars with 1 tsp of salt in each, adding hot

broth and leaving 1 ¼-inch headspace.
- **Raw Pack:** Pack the jars loosely with raw meat and 1 tsp salt, leaving 1 ¼-inch headspace. No liquid is required.
- **Process:** process for about 65 to 90 minutes, adjusting for altitude if needed.

Canning Chopped/Ground Meat:
- Chopped meat should be shaped into balls or patties and cooked until light brown.
- Sauté ground meat without shaping it.
- Drain off any excess fat.
- Fill the jars and add boiling water, tomato sauce, or meat broth up to 1 inch from the top. You can add salt if required.
- Process for 75 to 90 minutes, adjusting for altitude if required.

Chunks, Strips, or Cubes:
- Remove large bones and precook the meat until it is rare.
- Fill the jars, and add boiling water, meat broth, tomato juice, or meat drippings up to an inch from the top.
- Process for up to 75 to 90 minutes, adjusting for altitude if needed.

Preserving Dairy and Eggs

There are a few ways to preserve dairy and eggs, although most people are surprised to learn that eggs can be preserved.

Eggs:
- **Cold Storage:** unwashed eggs can be refrigerated or stored somewhere cold for months, so long as they still have the bloom on.
- **Freezer:** whip the eggs and freeze them in silicon cups or ice cube trays before transferring them to a bag. You can use these for some baking recipes, but they don't work well for scrambled eggs.
- **Liming:** limed eggs will be kept for at least a year. Use an ounce of pickling lime per half-gallon jar, adding a quart of water and whisking it together. Put the eggs in, pointed end down, and tighten on the lid and screw band to stop air from getting in.

Store in a dark, cool place.
- **Freeze-Drying:** You'll need a freeze-drier for this. Blend raw eggs and pour them onto the trays. Process per manufacturer's instructions. You can also freeze-dry cooked eggs in the same way. Store in airtight jars somewhere cool, dry, and dark. You can do the same by freezing the eggs on trays in your freezer.

Preserving Milk

Refrigeration is the obvious way to preserve milk, but it will only last a week or two before it is past its best. Other ways include:
- **Culturing:** make yogurt or kefir from it.
- **Cheese:** this is one of the best ways to use and preserve milk. You can make all kinds of cheese, but be aware that soft cheeses won't keep as long as the harder ones.
- **Freeze:** milk can be stored in plastic containers for a few months, but it will lose its texture over time, and fat globules will make it grainy when you thaw it.
- **Freeze-Drying:** use your freeze-drier as you do for the whipped eggs.

Bonus Chapter: Your Homesteading Checklist

Starting a homestead isn't easy, but it is incredibly rewarding. To help you make sure you do everything right, here's a checklist to follow:

STEP	MILESTONES	DONE
Set Your Budget	• Determine how much you can afford to spend on land/property. • Determine additional fees, including taxes. • Factor in building costs, including buildings, waste management, fencing, etc. • Factor in energy costs. • Can you afford to do this? If yes, move on.	

STEP	MILESTONES	DONE
Find the Land	• Do you want to live off-grid? • Find the right piece of land. • Consider the climate, land, labor costs, regulations and laws, location, proximity to amenities, soil quality, utilities, access, etc.	
Build or Renovate	• If there is already a house, consider renovation costs. • If not, consider building costs – DIY, prefab, or hiring a contractor. • You also need to consider extra buildings for storage and animals.	
Waste Management	• Design and build a waste management system to handle wastewater and sewage. Most people opt for a septic tank. • Get the required permits.	
Energy Requirements	• Are you hooked up to the grid, or do you want to off-grid? • Work out your power needs. • Determine whether you want solar, wind, hydropower, or a combination. • Hire the right contractors to ensure the job is done properly and with the required permits.	

STEP	MILESTONES	DONE
Plan your Homestead	• Visualize what you want to do with your homestead. • Draw a plan of your land and determine where your vegetable garden, fruit trees, chickens, beehives, and livestock will go.	
Prepare the land	• Check and amend the soil quality where needed. • Till the land to get it ready for planting, removing rocks and other debris. • If you don't have enough land for an extensive vegetable garden, build raised beds and plan for square-foot gardening.	
Build a Compost Area	• Find a small corner of the land you can use for compost. • Decide between a heap on the ground or a purpose-built compost heap. • Start filling it with weeds, vegetation, cardboard, and some food scraps to get it going.	

STEP	MILESTONES	DONE
Learn to Preserve Food	• Learn about the different types of canning. • Set up a cool room or root cellar to store veggies, fruits, cheese, butter, etc. • Learn how to freeze, dehydrate, smoke, and cure food to preserve it.	
Determine if You Want Livestock	• Have you got space? • Determine what you want to keep. Most poultry, rabbits, and bees don't take up much space. • If you have acres of land, decide if you want larger livestock, like cattle, pigs, goats, etc. • Ensure you can build the relevant shelters and have strong fencing. • Make sure you are allowed to keep livestock in your area. • Be sure you are willing to do what it takes to care for the animals and keep them healthy and happy.	
Find a Community	• Find a homesteader's community and befriend them. • Look local first and reach out to ask questions or get advice. • Join online communities to trade advice and tips.	

Perhaps the most important thing is to have the right mindset. You will fall at the first hurdle if you are not mentally prepared for the long haul.

Homesteading isn't easy, and you need a great deal of determination and dedication to succeed, not to mention the hard work that lies ahead. You don't need to do everything in one go. Take it one step at a time, start small, and plan to expand as you go along – a little every year will soon see your homestead build into a thriving place that you really enjoy being in.

Conclusion

Thank you for taking the time to download and read *"Country Living: The Ultimate Guide to Homesteading, Beekeeping, Raising Livestock, and Achieving Self-Sufficiency in the Countryside."* We hope you found it useful and now have a good idea of whether country living or homesteading is the life you want to lead.

Homesteading is what you make of it. You can go all-in and live off-grid completely, or you can simply cut your reliance on public utilities by using some renewable energy and water sources. You can become totally reliant on the food you grow and raise, or you can simply become less reliant on grocery stores.

This is your life. This is you, living the way you want to live, no matter how hard it is to start with. And make no mistake. This is not an easy life, at least not to start with. It will cost you financially but also in blood, sweat, and tears, but the rewards will eventually far outweigh what it costs you, and you'll end up living a life you fully enjoy.

Thank you once again for reading this book, and if you found it enjoyable, please leave a review on Amazon for other potential readers.

Part 2: Natural Parasite Management for Livestock

Harnessing Nature's Solutions for Internal Control and Healthy Herds

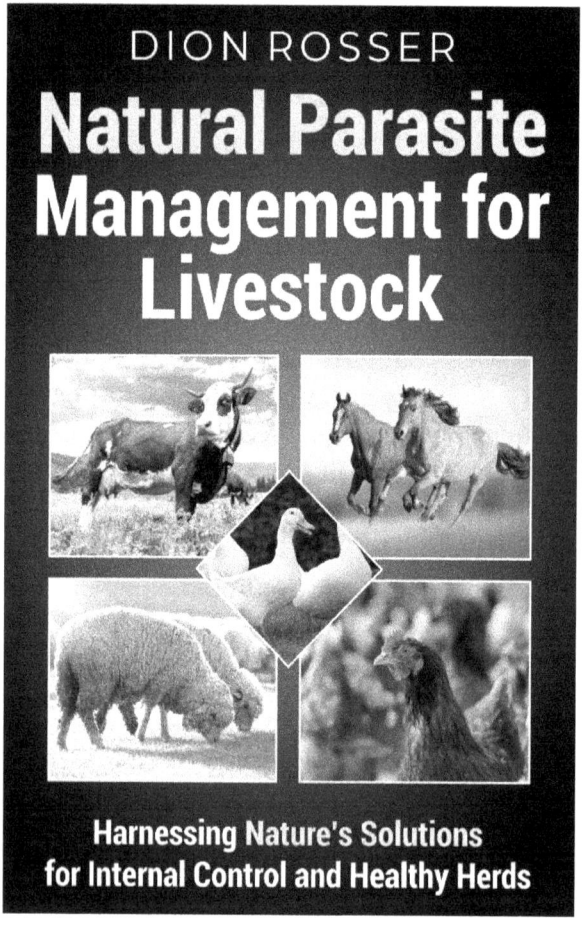

Introduction

As a livestock owner, you're probably no stranger to the daily rituals of tending to your animals, ensuring they're well-fed, comfortable, and healthy. But have you ever noticed those subtle signs that something might not be quite right? Perhaps your chickens' combs have lost their vibrant red hue, or you've noticed a horse or cow looking a bit more "ribby" than usual. Even your sheep's once rosy eyelids now appear paler than they should be. It's easy to dismiss these small changes as part of the natural ebb and flow of livestock care, but these indicators may be telling you something crucial – your animals might be under attack from parasites.

Parasites can be silent predators, slowly sapping the vitality and well-being of your livestock. They lurk beneath the surface, affecting your animals in ways that may not be immediately evident. The reality is that managing parasites can become a relentless, full-time job, especially if you're committed to doing it without resorting to chemical solutions. But what if there was a way to shift this burden from an all-consuming endeavor to a part-time, seasonal one that you and your livestock could benefit from?

A pale comb and wattles may be the first indication of a possible parasite problem for chickens. Rest assured, though, by developing a systematic approach, you can transform the daunting task of parasite control into a manageable and sustainable practice. This transformation is achievable through several key strategies:

- **Pasture Management and Rotation**: You must learn to optimize your pasture management techniques and rotation practices to

minimize the risk of parasitic infestations. Careful management of grazing areas can reduce exposure to parasites, giving your livestock more than a fighting chance to flourish.
- **Environmental Control**: Learn to implement effective measures to create an environment less conducive to parasites. By making your livestock's living conditions less hospitable for these intruders, you can reduce the prevalence of parasitic infections.
- **Targeted Deworming with Natural Products**: Embrace the power of natural solutions for deworming, possibly even growing some of the preventative ingredients yourself. Discover how to use targeted deworming strategies with organic and sustainable products, ensuring that you only treat when necessary, minimizing the risk of resistance.
- **Continual Research and Learning**: Stay up-to-date with the latest developments in parasite control for livestock. An ongoing commitment to research and education is essential in adapting to new challenges and optimizing your parasite management strategies.

A healthy and vibrant livestock operation is within your reach. Healthy chickens should have bright combs, shiny feathers, and lively spirits, while cows, horses, and sheep should showcase the robustness that's the hallmark of well-cared-for animals. While specific symptoms are associated with particular parasites, it's crucial to remember that these indicators are not definitive diagnoses. High parasitic loads can lead to various issues, including diarrhea, dehydration, weight loss, and lethargy, regardless of the specific parasite responsible. But don't worry; this book is here to guide you.

As you start this journey to reclaim the health of your livestock, remember that you're not alone. There is a world of resources out there, starting with this comprehensive book! So, dive right into the world of sustainable and holistic livestock care.

Chapter 1: What Are Livestock Parasites?

To give you a foundation for the information you'll learn from the book, this chapter explores various types of parasites that affect livestock, their groups, background, life cycles, and their potential impact on animal health and productivity. It also emphasizes the importance of understanding and managing these parasites in livestock production.

History and Background of Livestock Parasites

Parasites are living beings living inside or on the surface of another organism (known as a host), taking nutrients from the latter. Livestock and other animals can be affected by over 1000 parasite species, some of which can also be transmitted to humans.

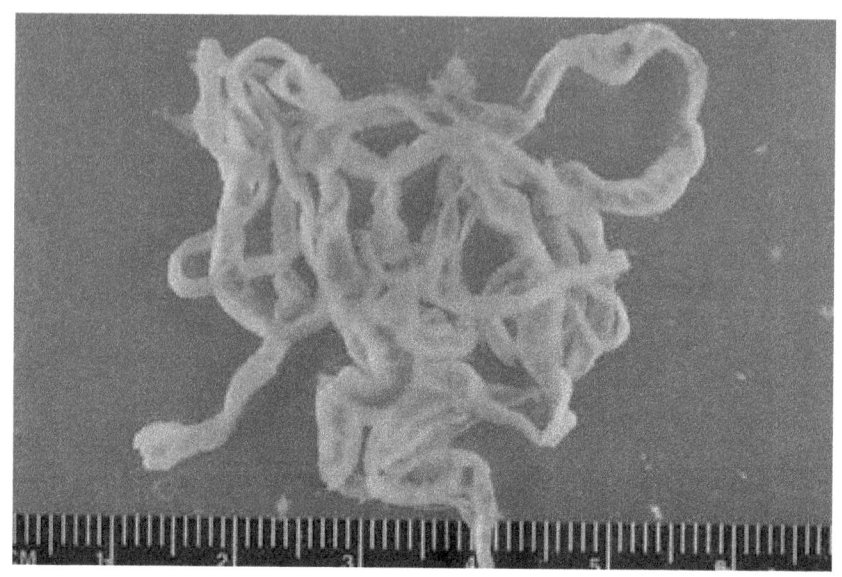

Parasites are living beings living inside or on the surface of another organism (known as a host), taking nutrients from the latter.
FWC Fish and Wildlife Research Institute, CC BY-NC-ND 2.0 DEED
<https://creativecommons.org/licenses/by-nc-nd/2.0/>https://www.flickr.com/photos/myfwc/14837075886

Parasites, in a veterinary study field called parasitology, have their roots in ancient civilizations. Ancient Egyptian records suggest that Egyptians studied parasites and described larger ectoparasites (or external parasites) in humans but didn't understand the organism's life cycle. Similar parasites are believed to have affected the Israelites during their voyages; they described them as fiery serpents. The first to recognize that parasites had several stages during their life cycle was Aristotle, who noted cysts of worms in pigs' tongues. Scientists also theorize that the Hebrews prohibited pig consumption because they likely discovered similar cysts in these animals.

From the first century onward, interest in parasites grew intensely. At the beginning of the second century CE, Aretaeus noted finding several fluid-filled bladders in animals, and a century later, Galen described three distinct types of parasites in humans. In the 7th century, Paulus Aegineta studied human helminths even more, naming one of the groups Ascarides. These small worms, located in the lower intestines of people and warm-blooded animals, are the group of tapeworms and ascarid worms now known as Ascaris.

A Byzantine physician named Alexander authored the first book about parasitic worms, *De Lombrices* (translated as "on worms"), establishing the foundation of modern parasitology. However, at this time, scientists and scholars only knew that parasites cause diseases, but not how they came to the host and how to prevent this. The source of parasitic infection was identified at the beginning of the 11th century when Ibn Zuhr (an Islamic Moroccan physician) and Abbess Hildegard of Bingen (an animal health writer and researcher) both concluded that mites transmitted scabies.

Parasitology understanding was expanded during the Middle Ages when Albertus Magnus wrote about parasitic worms (helminths in contemporary literature) in fish, horses, falcons, and dogs in his book *De Animalibus* in 1478. A couple of decades later, Anthony Fitzherbert described the disease that was caused by the liver fluke in his work *A Newe Treate or Treatise Most Profytable for All Husbandmen*. He even concluded that the source of the infection was wet, marshy land where snails also lived, but he didn't understand the parasites' life cycle enough to make a connection between the snails and the infected animals. In his first classification of animals, Linnaeus wrote about fasciola hepatica, described as a leech whose young thrive in water. Because the parasitic cycle was not understood until much later, people believed that parasites are spontaneously generated in the bodies of people and animals.

William Harvey's discovery of the heart's role in blood circulation was one of the earliest discoveries that cast doubt on the hypothesis of spontaneous genesis. In addition to describing this theory in his dissertation On Animal Generation (published in 1651), Harvey contends that all living things originate from eggs instead of spontaneously emerging. Dutch biologist Jan Swammerdam described several life forms of insects, including adult, pupa, larva, and egg, proving that these animals go through an entire cycle during their lifetime, proving that they didn't come from nowhere.

In the 17th century, Italian physician Francesco Redi noted, removed, and examined ticks and lice from people and animals. Redi described the "louse" as one of the diseases these parasites caused, for which he became known as the "father of parasitology." Through a simple experiment, he also once and for all proved the theory of accidental generation invalid. Laying two cuts of meat on a plate, he covered one, leaving the other one uncovered. The latter soon attracted flies, which laid eggs on it, and in two days, the meat was infested with maggots.

In contrast, the covered piece of meat did not have any maggots in it. Confirming Redi's findings, Dutch microscopist Antonie van Leeuwenhoek observed protozoan parasites under the microscope and sketched them. These were parasites found in peoples' and animals' intestines, often causing diarrhea. English physician Edward Tyson studied and examined the nematode Ascaris lumbricoides, ultimately finding that the parasite had two sexes (which alludes to sexual reproduction) and further disproving the theory of random generation in parasitic worms.

Numerous animal and human parasites were found and characterized during the 17th and 18th centuries. Johann Goeze documented the ascaris worms in pigs, while Peter Simon Pallas reported the hydatid cysts in people and the cat tapeworm (taenia crassiceps) in 1766. Swedish and German botanists published three tomes systemizing parasite species in 1819, establishing a standard reference that was considered valid until the parasite life cycle was fully understood. In 1863, German physician and pathologist Rudolph Virchow suggested that a more vigorous pig meat infection could prevent trichinosis infection in people.

After discovering a previously unidentified worm in the bile duct of a giraffe, British doctor TC Cobbold set out to study and perfect the current parasitic systemization. In 1878, he presented the discovery of a filaria embryo in a mosquito's body, which led to the concept that ties mosquitos to the disease known as malaria. Two years later, Griffith Evans discovered that the first pathogenic trypanosome, Trypanosoma evansi, was the cause of a tropical illness that affected horses and camels.

With these and similar discoveries in the mid to late 19th century, parasitology became a well-established field of study in veterinary medicine. Scientists also began to look into the parasite life cycles even more, which allowed them to come up with effective control, prevention, and treatment measures for the diseases they caused. People recognized that some parasites, like trichinella and other worms, represented severe public health hazards and had a great need to control their spread. By investigating these and other parasites affecting livestock, control measures for liver fluke, lungworms, trypanosomiasis, coccidia, haemonchosis, roundworms, and other parasites were developed and successfully implemented.

Why Is Parasite Control Important in Animal Husbandry?

Both external and internal parasites can cause health issues in animals, weakening their immune system and predisposing them to bacterial and other infections. They might also cause damage on their own depending on how they feed on the host and whether they carry and transmit other diseases. Infected livestock can also transmit the parasites or the secondary diseases they carry to humans and other animals. All these led to significant economic losses, just as they did since people started domesticating animals and rearing them for food and other purposes. For all these reasons, controlling parasitic proliferation in animal husbandry is of utmost importance. Fortunately, due to all the discoveries of parasitology nowadays, there are numerous measures for control and prevention.

Common Livestock Parasites

Livestock parasites are divided into two major categories – internal and external. The internal ones enter the animal's body, feed on it, and damage it from the inside, while the external ones attack and infect from the outside. Based on whether they are transmitted to animals from another source, parasites can also be vector or non-vector-borne. Vector-borne organisms are typically internal as well.

Internal Parasites and Diseases
Coccidia

Coccidiosis is caused by the protozoan Eimeria sp., commonly known as coccidia. They live in farm animals' intestinal walls and are host-specific. In other words, the variant that infects cattle will not infect goats and sheep. While coccidia is normally present in animals, over-proliferation can cause infestation and diseases. The latter typically happens in young specimens lacking sufficient immunity (premature weaning, cold climate, etc.) to prevent proliferation or in older animals in overcrowded or stressful conditions. Coccidia is transmitted by feces, which can remain infectious on the ground for up to six days and spreads very quickly in warm and wet conditions.

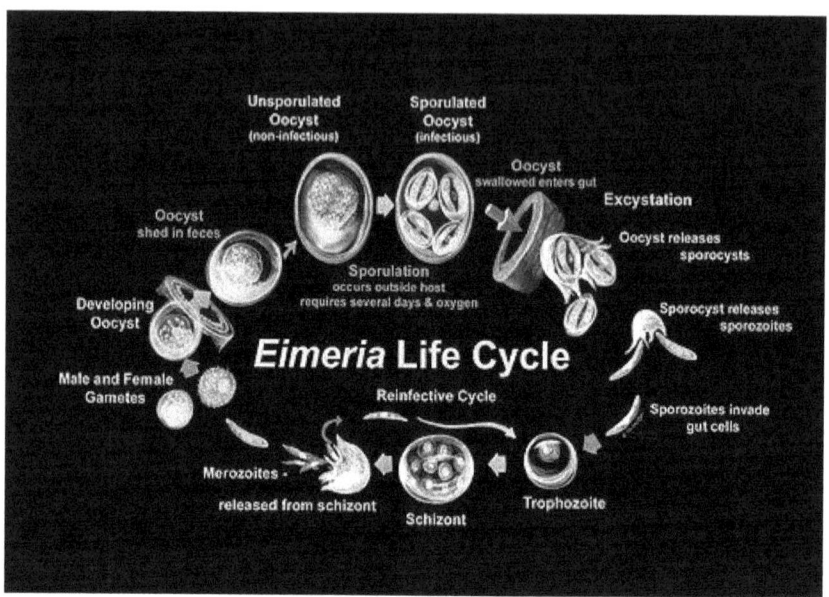

While coccidia is normally present in animals, over-proliferation can cause infestation and diseases.
https://commons.wikimedia.org/wiki/File:Eimeria_life_cycle_usda.jpg

Nematodes – Gastrointestinal Worms

Most worms in the gastrointestinal (GIT) tract of livestock are nematodes. While this specific group of worms has several common features, not all prefer the same conditions (for example, some prefer warmer climates, while others thrive in cold weather). Young animals are at a higher risk, but GIT worms can affect other specimens, too, if those have lower immunity to intestinal parasites (like bulls, for example). Expelled from the GIT tract, parasitic nematodes can survive on the ground for several days – the exact length depends on the climate conditions and whether they have a nutrient source available.

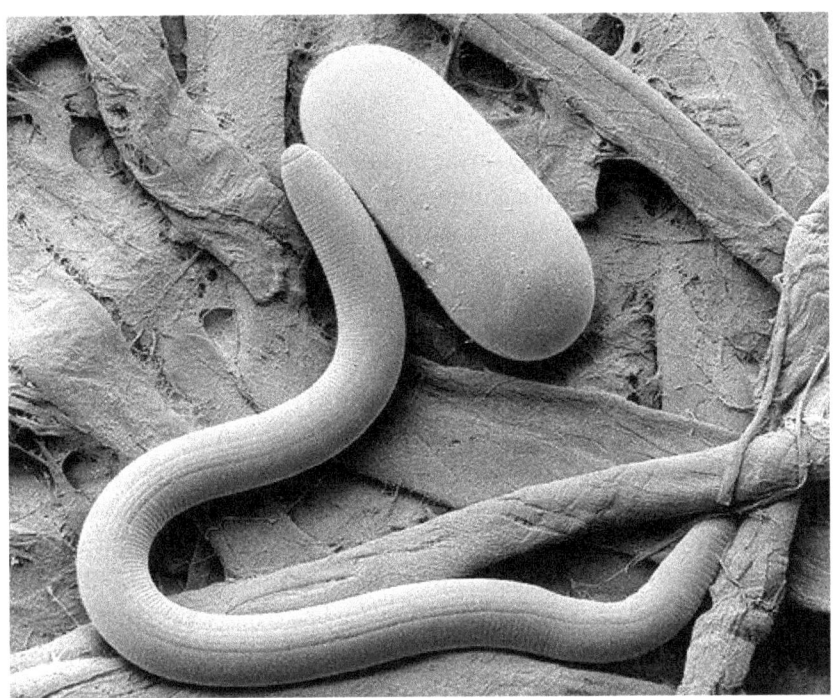

Expelled from the GIT tract, parasitic nematodes can survive on the ground for several days.
https://commons.wikimedia.org/wiki/File:Soybean_cyst_nematode_and_egg_SEM.jpg

Liver Fluke

Caused by the parasitic trematodes platyhelminths, liver fluke is a zoonotic disease that, while common in livestock, can also infect other animals and people. The worms live in the liver and bile ducts and can't survive on the ground for more than two days. However, they have another host, snails, making areas like marshes, springs, coastal regions, irrigated pastures, water troughs, etc., potential sources of infection.

Theileriosis

Theileriosis is a vector-borne disease transmitted by ticks and, on rare occasions, by other biting animals or reused veterinary injection needles. It's brought on by theileria orientalis, a blood parasite causing anemia. The parasite can survive on surface food for 12-24 hours, depending on conditions.

Trichomoniasis

Caused by the protozoan parasite tritrichomonas foetus, trichomoniasis is a venereal disease that lives in the host's genital tract, often leading to loss of embryo, abortion, and stillbirth in female animals. It is transmitted via mating or insemination, but the parasite can survive on non-live surfaces for up to 24 hours.

Toxoplasmosis

Caused by the protozoan parasite Toxoplasma gondii, toxoplasmosis is another zoonotic disease disseminated by infected feces or, in some cases, by rodent bites. It can also be transmitted by animals eating infected tissue of other animals. The parasite lives for several days in a cyst or asexual reproductive form in the dung and other surfaces.

External Parasites and Diseases

Flies are common livestock parasites, and depending on the species, they can cause issues on their own or transmit vector-borne diseases. For example, buffalo flies bite and feed on the animal's blood by latching onto its skin and laying eggs on its dung. They can travel over six miles in search of a host. Animals in poor condition and those with darker coatings are more likely to attract flies. The same animals are more at risk of contracting flystrike, caused by blowflies laying their eggs on the animals (in wounds, unclean mucous surfaces like nose, genitalia, etc.). Flystrike is an extremely painful and often fatal condition. Nuisance flies, on the other hand, only feed and breed in the dung – however, for the same reason, they carry diseases, which, when transmitted from feces, can infect livestock.

Ticks

While ticks often have specific hosts (for example, cattle ticks will only feed on and infect cattle), they can also survive on other animals and humans. Female ticks drop off their eggs on the same animal they feed on. The larvae hatch, develop in nymphs, and become adults within 21 days, often remaining attached and feeding on the same animal. Ticks are carriers of vector-borne diseases, and some species will cause health issues on their own.

Tick Fever

Caused by blood parasites and transmitted by ticks, tick fever is a serious disease that destroys the host's blood cells, affecting several organs and often leading to death.

Tick Paralysis

Ixodes holocyclus is a tick species that secrete a toxin in its saliva, which causes paralysis in animals and people. Small farm animals are more vulnerable than larger livestock due to the amount of toxin per body mass ratio. This is a three-host tick. Each phase has its host, but only adults can cause paralysis. The larvae and nymph stages are vulnerable and can survive only a few hours without a host. The same applies to bush ticks, which transmit the blood parasite Theileria orientalis to warm-blooded animals.

Lice

Livestock is vulnerable to both sucking and biting lice, both host-specific external parasites. Biting lice feeds (bovicola sp.) latch onto the nasal skin and feed on the skin and the bacteria that grows there. Sucking lice (linognathus sp.), on the other hand, have a specific mouthpiece that can penetrate the animal's skin, enabling them to feed on its blood.

Livestock Parasites Life Cycles and Control Measures

Here are the life cycles of common livestock parasites and the stages where control measures should be implemented.

Nematodes

With a few exceptions, most nematodes have a similar life cycle, which is as follows:

1. The adult female nematodes lay their eggs in the animal's GIT.
2. The eggs are expelled into the dung.
3. While in the dung, the eggs develop into first-stage larvae, then molt to second-stage larvae (since they are less mobile, the larvae are more vulnerable to antiparasitic measures).
4. Feeding on the animals' dung, the larvae reach the second molt stage and develop into third-stage larvae.
5. The third-stage larvae are more mobile and migrate to the vegetation where the animals graze and ingest them (before ingestion, the infection can still be controlled by preventative grazing measures).
6. Once ingested, the third-stage larvae develop into fourth-stage larvae in two to five days.
7. After 14 days, the fourth-stage larvae become adult nematodes that can live up to several months, feeding and reproducing.

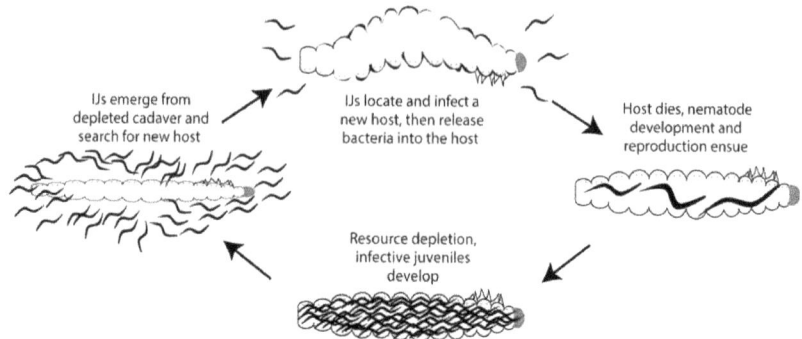

With a few exceptions, most nematodes have a similar life cycle.
Adler Dillman, CC BY 4.0 <https://creativecommons.org/licenses/by/4.0>, via Wikimedia Commons: https://commons.wikimedia.org/wiki/File:EPN_Lifecycle.tif

Toxoplasma

Toxoplasma has livestock as its definitive host for both asexual and sexual reproduction. Depending on the stage upon ingestion, the toxoplasma life cycle can last between a few days (for bradyzoites) and three weeks (for tachyzoites). The full lifecycle is as follows:

1. After the process of sexual reproduction, the parasites produce oocysts, which are expelled from the animal body through feces a few weeks after the infection – this is the time when control measures are the most effective as they can prevent the rest of the cycle.

2. The oocysts contain sporocysts, and within a few days, four sporozoites develop within the sporocysts, beginning the sporulation process (asexual reproduction) –the fastest phase of reproduction.

3. When another animal ingests the sporulated oocysts (bradyzoites), the sporozoites exit the oocyst and invade the animal's small intestine, where they enter the enterocytes (intestinal cells) – the pace of reproduction slows down.

4. Alternatively, the sporozoites can invade the host's blood and lymph cells (by going through the intestinal wall) and become tachyzoites.

5. As tachyzoites, they reach the tissues, where they develop into bradyzoites contained in tissue cysts in cardiac and skeletal muscle, eyes, and nervous system tissues – control and antiparasitic measures can be implemented here to kill the cysts and boost the animal's immunity to prevent cyst development.

6. Tachyzoites can also be expelled if they reach the intestines (and consequently be ingested by other animals).

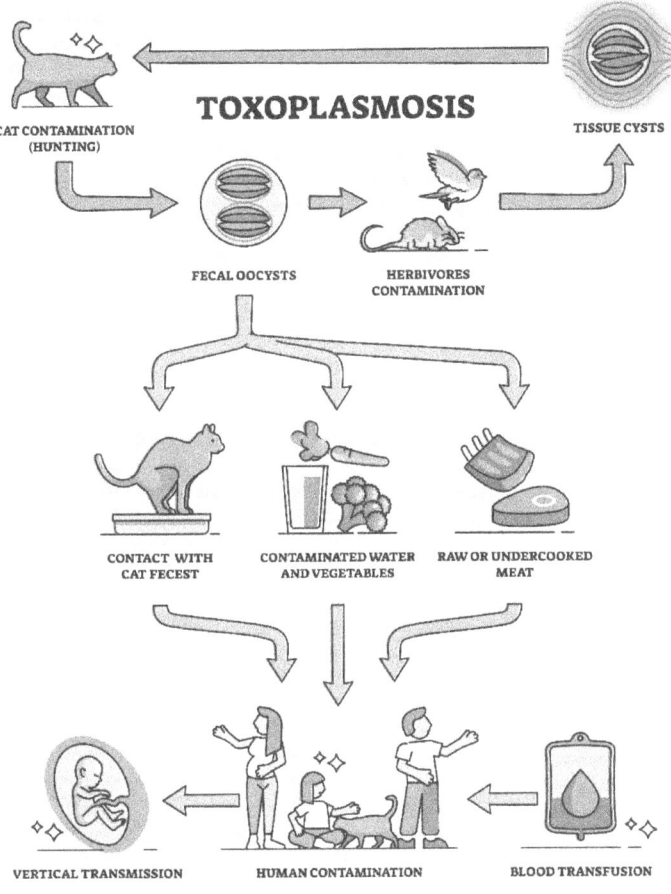

Toxoplasma has livestock as its definitive host for both asexual and sexual reproduction.

Liver Fluke

As a two-host parasite, liver fluke has snails and livestock as its first and definitive hosts. The full cycle is as follows:

1. The parasite produces eggs in the snail, where they develop into first-stage larvae – without the interaction of the host, the cycle can't be completed, so preventing it is a crucial control and antiparasitic measure.
2. The first phase larvae exit the snails and get to the herbage, forming cysts.

3. The grazing animals consume the cysts, which break, and the larvae continue to develop.
4. The larvae get through their intestinal wall and into the liver, where they become adults feeding on the bile duct.
5. Through the bile, the adult worms get into the feces, producing eggs expelled into the dung.
6. The eggs are consumed by snails, where the cycle continues (the entire cycle lasts up to 21 days.

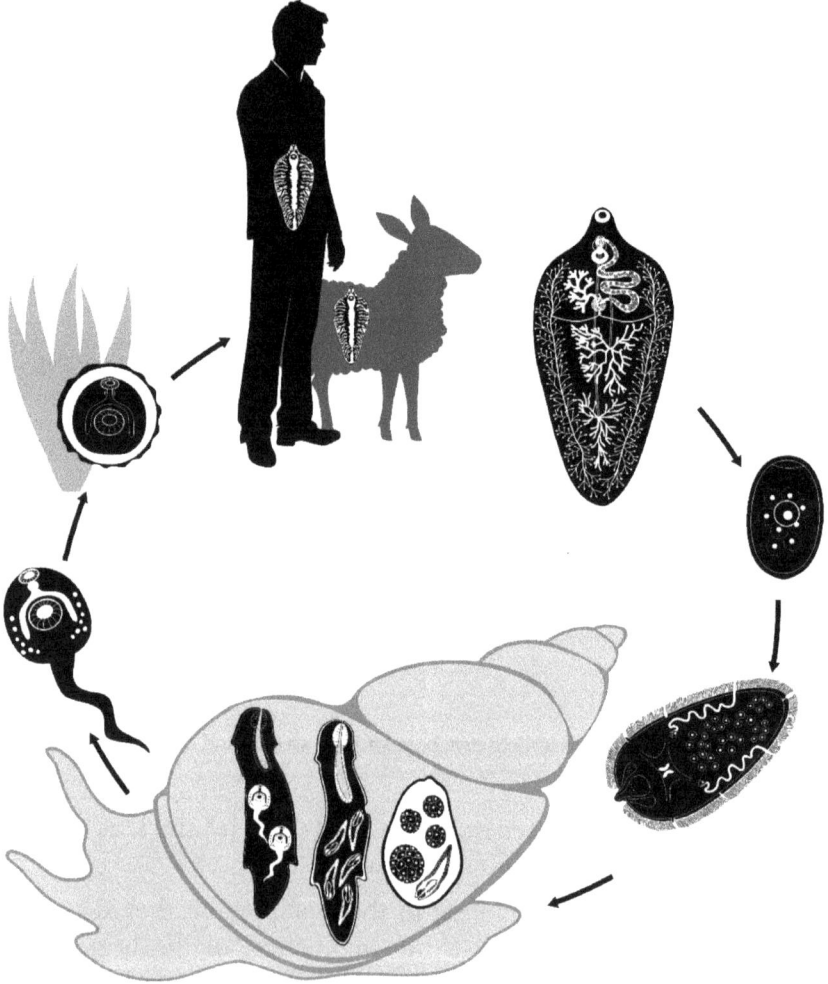

As a two-host parasite, liver fluke has snails and livestock as its first and definitive hosts.

The Negative Effects of Parasitic Infections

Livestock parasites have numerous adverse effects which impact animal welfare and cause economic losses for farmers. Below are the negative effects of parasitic infections in livestock.

Impaired Liveweight Gain

Feeding in the host's intestines, GIT parasites divert nutrients from the host, which results in its inability to gain weight. However, external parasites and vector-borne diseases can also have the same effect as they cause circulatory and gastrointestinal issues and reduced appetite. Depending on the type of animals in question and the degree of parasitic infection, weight gain can be reduced by up to 90 pounds.

Effect on Milk Production

Certain parasites affecting the internal organs could also cause reduced milk production in lactating animals. This is a particularly economically devastating effect for animals raised for their milk. Younger female specimens (those in their first or second lactation period) are more affected by intensive parasitic infections and are more at risk of having reduced milk production.

Reduced Meat Quality

Since many livestock animals are raised for their meat, parasite infection reducing carcass quality can also represent a significant issue. Nematodes are particularly problematic as they form cysts in the muscles, reducing their quality and hindering their growth. Instead of muscle, some of these animals will grow more subcutaneous fat, which isn't as profitable. Other parasites that cause reduced appetite, GIT issues, and limited blood and nutrient distribution through the body can have similar effects.

Impaired Reproductive Performance

Certain parasites can lead to lowered conception rate and breeding ability in livestock. This is tied to the failure to gain weight and the type of parasitic infection. For example, toxoplasmosis is a commonly known cause of infertility in young livestock. The likelihood of conception and regular breeding intervals increases if the parasitic infection is cured before animals reach breeding weight. On the other hand, GIT parasites rarely cause infertility issues or hinder reproductive performance in livestock.

Increased Mortality Rate

Parasitic infection can increase livestock mortality rates for two reasons. If the parasites cause symptoms that lead to the animal's failure to flourish and inability to heal, they can lead to the animal's death. For example, certain GIT worms can lead to a 100% mortality rate in animals aged one to three months. The young specimens have a smaller intestinal tract, which the worms obstruct, preventing the animal from taking in any nutrients. In adult livestock, this rarely happens because their GIT track is larger. Besides the immediate cause, parasites can also be an indirect cause of livestock mortality because they compromise the immune system, so the animals succumb to other, far more serious conditions. For instance, if a worm larva or cyst migrates to the lungs, this can cause pneumonia or toxemia, resulting from the organism's inability to fight off infections.

Effects on Public Health

Certain parasitic infections (like Trichinos or tapeworms) can harm public health. These parasites can infect humans and other animals if inadequate control measures aren't implemented. Treating these secondary infections further strains the economic burden of parasite management.

How Seasonal Variations Affect the Prevalence of Parasitic Infections

Seasonal variations in weather and climate can influence the prevalence and intensity of parasitic infections. For example, most protozoans show a greater presence in the warmer months (May to September). By contrast, nematodes are more likely to cause infections in winter (December to February). This is likely because free livestock breeding is more frequent from spring through autumn, which creates ideal conditions for protozoan infections. While grazing in open pastures, livestock is more exposed to external parasites and can also ingest internal ones. During the winter months, the animals are often kept in a closed space, and their nutrition is controlled (they have lower chances of ingesting parasites), which lowers the prevalence of some parasitic infections. Moreover, protozoans are more affected by seasonal changes and changes in temperature on a day-to-day basis, while nematodes aren't.

If the animals are kept in a closed space during the winter and gather around a small feeding area, this further increases the chances of parasitic infections caused by nematodes. During the warmer months (and in warm, enclosed spaces), nematode larvae can move quickly and will consume more nutrients. When nutrients are scarce, the larvae will stop moving to conserve energy. If they can't find more nutrients, they'll die. The increased water loss during the summer further accelerates this process. In colder climates and seasons, the larvae won't move as quickly because they require more energy to do that. They can, however, curl up and form a protective cyst, which allows them to survive for up to several months, depending on the species and climate conditions.

Some worms, like the brown stomach worm or the black scour worm, prefer wet and cold conditions, especially in moderate climates, where the winters are characterized by higher temperatures and heavy rainfalls. Other worms, such as the barber's pole worm, for example, prefer warm conditions and regions with heavy rainfall or increased irrigation. If livestock graze pastures with large amounts of precipitation (natural or artificial) during the summer, they'll be more likely exposed to this worm.

Fly larvae prefer warm conditions, so the prevalence of flies and fly strikes is higher during the warmer months than during the winter. The risk of flystrike increases during the spring as soon as temperatures reach above 63 degrees Fahrenheit. Rainfalls and moderate wind conditions in spring through autumn further increase the prevalence of flies and fly larva development in manure, natural vegetation, and other places where flies are attracted (for example, where urine accumulates).

Interestingly enough, lice have no preference for climate conditions and can proliferate just as easily during the winter as they do during the warmer months. Like flies, lice eggs and larvae will survive in vegetation and on the surfaces and animals in an enclosed space. Sucking lice are more likely to cause a parasitic infestation during the winter when the animals are kept inside.

Chapter 2: Clinical Signs and Diagnosis

Parasitic infections in livestock can have devastating economic and health implications for farmers and ranchers. These outbreaks and infections lead to reduced productivity, weight loss, and, in severe cases, mortality. To manage this situation effectively, it is essential to recognize and diagnose these conditions accurately. This chapter explores common parasitic infections in each livestock species, focusing on identifying and diagnosing them effectively.

Parasites and Livestock

Whether it's a small farm with a few animals or an industrial-scale livestock farming operation, these infections, when left unattended, result in many challenges, including economic losses, animal suffering, and environmental consequences. To develop a clear perspective, here's why timely recognition and diagnosis of parasitic infections in livestock is necessary.

Animal Welfare

On-time recognition and diagnosis of parasite infection or outbreak is crucial to establish animal health and welfare. The discomfort, pain, and suffering livestock animals experience can be alleviated through early intervention. These parasitic infections, if left unattended, can even result in severe conditions that lead to organ damage and, in some cases, even death.

Economic Impact

No matter the number of animals you care for, it's necessary for livestock farming to be economically sustainable. Parasitic infections and diseases will decrease productivity as animals face issues like weight loss, reduced milk or egg production, and lowered reproductive performance. All these factors can have a severe economic impact if not handled properly.

Prevention of Parasite Spread

When you diagnose parasitic infections early, you'll be in a good position to prevent them from spreading by quarantining infected animals quickly. This proactive approach helps to protect the rest of the population from potential outbreaks and minimizes the need for extensive treatments.

Targeted Treatment

Accurate diagnosis enables veterinarians and livestock owners to develop customized treatment strategies. This tailored approach reduces the unnecessary use of medication and minimizes the risk of drug resistance. By focusing on treating only infected animals, farmers can achieve cost savings.

Environmental Impact

Recognizing and diagnosing parasitic infections also has positive environmental implications. Treating infected animals minimizes the spread of parasites through their feces, which benefits the local ecosystem and other livestock sharing the same environment.

Food Safety

Parasitic infections in livestock can affect the safety and quality of meat, milk, and eggs. Through accurate diagnosis and subsequent treatment, livestock owners can ensure that these products meet safety and quality standards, safeguarding the health of consumers and preserving the reputation of farming operations.

Research and Epidemiology

Accurate diagnosis of parasitic infections offers valuable data for research and epidemiological studies. This information aids scientists in understanding the prevalence and distribution of parasitic infections in different livestock populations. Such research forms the basis for developing more effective control and prevention strategies, ultimately benefiting the livestock industry.

In various regions, legal requirements exist for controlling and treating infections in livestock. Proper diagnosis and adherence to these regulations are essential to prevent potential legal consequences, ensuring compliance with local laws and regulations.

Common Clinical Signs

Clinical signs of various animal health conditions can be categorized based on the affected organ systems or body parts. Here are some common clinical signs you can find in different organ systems.

Gastrointestinal Signs:
- **Diarrhea:** Common in gastrointestinal parasitic infections, diarrhea can vary in severity and may contain mucus or blood.
- **Weight Loss:** Parasites can lead to chronic weight loss due to reduced nutrient absorption.
- **Decreased Appetite:** Infected animals may eat less, resulting in malnutrition.
- **Bottle Jaw (Edema of the Lower Jaw):** Swelling under the jaw due to edema and anemia, often seen in small ruminants with heavy gastrointestinal worm burdens.
- **Dehydration:** Persistent diarrhea can cause dehydration, sunken eyes, dry mucous membranes, and decreased skin elasticity.
- **Abdominal Pain:** Some animals may exhibit colic or discomfort in response to gastrointestinal parasites.
- **Submandibular Edema (Brisket Disease):** Swelling under the jaw or in the brisket area, common in cattle with high parasite loads.

Dermatological Signs:
- **Hair Loss (Alopecia):** Parasitic skin infections, such as mange mites, can lead to hair loss and skin irritation.

Parasitic skin infections, such as mange mites, can lead to hair loss and skin irritation.
Alan R Walker, CC BY-SA 3.0 <https://creativecommons.org/licenses/by-sa/3.0>, via Wikimedia Commons: https://commons.wikimedia.org/wiki/File:Sweating-sickness-Zimbabwe.jpg

- **Skin Lesions:** Skin conditions like dermatophytosis result in scabs and crusts on the skin.
- **Intense Scratching or Rubbing:** Animals may scratch or rub excessively due to itching caused by parasites, leading to hair loss and skin damage.
- **Pruritus:** Intense itching is a symptom of various skin parasites, causing animals to scratch and rub their skin raw.
- **Crusting and Scaling:** Affected areas may develop crusts, scales, or flakiness due to skin irritation.

Respiratory Signs:

- **Coughing:** Lungworm infections can lead to coughing and respiratory distress, particularly in cattle and small ruminants.
- **Nasal Discharge:** Seen in cases of lungworm and nasal bot infections, with mucus or nasal discharge from the nostrils.
- **Dyspnea:** Respiratory distress and difficult breathing can occur in severe cases of lungworm infection.

- **Crackles and Wheezes:** Abnormal lung sounds may be auscultated in animals with lungworm infections.

Ocular Signs:
- **Conjunctivitis:** In cattle and sheep, eye irritation and inflammation occur with certain eye-dwelling parasites, such as thelazia.

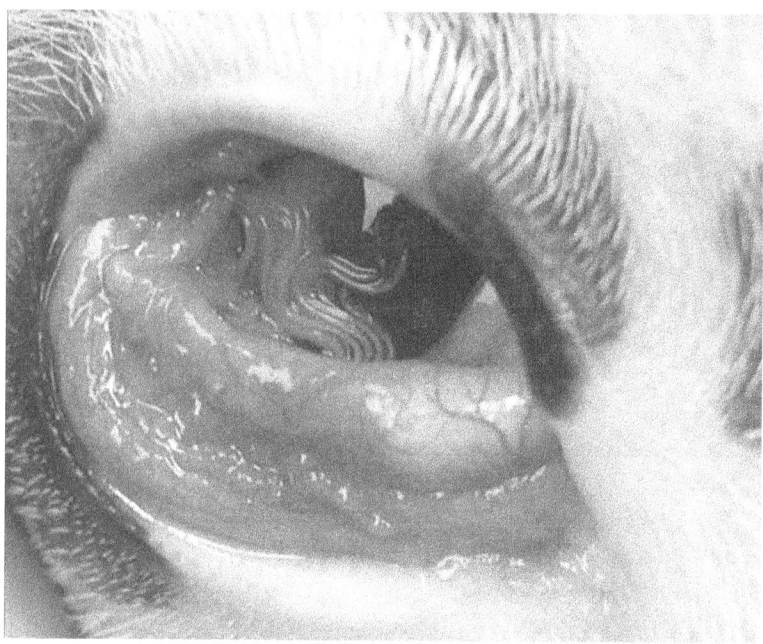

In cattle and sheep, eye irritation and inflammation occur with certain eye-dwelling parasites, such as thelazia.
https://commons.wikimedia.org/wiki/File:Thelazia_callipaeda_in_dog.jpg

- **Ocular Discharge:** Excessive tearing or discharge from the eyes may result from irritation by eye parasites or migration of larvae in the eyes.
- **Corneal Opacities:** Clouding or opacities in the cornea can be seen in cases of parasitic keratitis.
- **Corneal Ulcers:** Ulcerations on the cornea may result from parasitic eye infections, causing pain and discomfort.

Neurological Signs:
- **Circling or Head Tilt:** Signs of neural larval migrans caused by certain parasites, such as Baylisascaris, leading to incoordination and abnormal head movements.

- **Incoordination or Paralysis**: Infestations with certain parasites can affect the nervous system, causing incoordination and paralysis in affected animals.
- **Tremors or Seizures:** Some parasitic infections can lead to tremors or seizures, affecting muscle coordination.

Hematological Signs:

- **Anemia:** Blood-feeding parasites like ticks and haemonchus worms can lead to anemia, resulting in pale mucous membranes and weakness.
- **Pale Mucous Membranes:** Anemia, characterized by pale gums and eyes, is a common sign due to the loss of red blood cells.
- **Thrombocytopenia:** A decrease in platelet count results from certain parasitic infections, increasing the risk of bleeding disorders.

Urogenital Signs:

- **Urinary Tract Infections:** Certain parasites can affect the urinary system, leading to signs like frequent urination, painful urination, or discomfort.

Reproductive Signs:

- **Abortions:** Protozoal parasites like neospora and toxoplasma can lead to abortions in cattle, resulting in reproductive losses.
- **Prolonged Estrus (Heat):** Some parasitic infections can disrupt the estrous cycle, leading to prolonged or irregular periods of heat.
- **Reduced Libido:** Parasitic infections can lower libido in male animals, reducing mating activity.
- **Delayed Puberty:** Infestations with certain parasites can delay the onset of sexual maturity in young animals.

Gastric Signs:

- **Bloat (Ruminal Distension):** Common in cattle with gastrointestinal parasitic infections, parasites can interfere with normal digestive processes and cause bloat.
- **Gastric Ulcers:** Some parasites can lead to gastric ulcers, causing pain, decreased appetite, and weight loss in affected animals.

Hepatic Signs:
- **Liver Fluke Infection (Fasciola hepatica)**: In sheep and cattle, liver fluke infections can result in liver damage, leading to signs like jaundice (yellowing of mucous membranes and skin), unthriftiness, and an enlarged liver.
- **Ascites (Abdominal Fluid Accumulation)**: Liver fluke infections can lead to ascites, with a swollen, fluid-filled abdomen.

These clinical signs may vary depending on the parasite and the affected host species. Accurate diagnosis and treatment often require veterinary consultation, diagnostic testing, and management strategies specifically for parasites and livestock species.

Veterinarians will recognize these clinical signs and conduct thorough examinations to diagnose and treat underlying animal health conditions. Studying these signs leads to valuable clues and guides further diagnostic investigations.

Diagnosing Parasitic Infections

Although common clinical signs are mentioned earlier, in this section, you'll learn how to identify and diagnose them in livestock and identify examples of common parasitic infections in each species.

Common Parasitic Infections in Cattle

Gastrointestinal Nematodes:
- **Ostertagia Ostertagi (Brown Stomach Worm)**: This worm primarily affects the abomasum (the fourth stomach) and can cause clinical signs such as diarrhea, weight loss, and reduced feed intake. In severe cases, it can lead to anemia due to these blood-feeding parasites.
- **Cooperia Species**: These small intestinal worms cause diarrhea, poor weight gain, and suboptimal feed utilization.
- **Haemonchus Contortus (Barber Pole Worm)**: It's another blood-feeding worm that can cause severe anemia, pale mucous membranes, bottle jaw (swelling under the jaw), weight loss, and death if left untreated.
- **Liver Flukes (Fasciola Hepatica)**: Liver fluke infections can reduce milk production and trigger the development of jaundice (yellowing of mucous membranes) and weight loss. They

primarily affect the liver and bile ducts.
- **Lungworms (Dictyocaulus Viviparus):** Lungworm infections cause coughing, increased respiratory rate, and nasal discharge due to lung and airway damage.
- **Ticks and Mites:** External parasites like the common cattle tick (rhipicephalus (boophilus) microplus) cause skin irritation, restlessness, hair loss, and the transmission of diseases like anaplasmosis.

Recognition and Diagnosis:
- **Clinical Signs:** While it may be difficult for beginners to identify the signs, veterinarians and experienced livestock farmers can recognize parasitic infections based on observed clinical signs, including diarrhea, coughing, and skin lesions.
- **Fecal Egg Counts:** Examining fecal samples using techniques like the McMaster method to help identify the type and quantity of internal worm eggs, aiding in deworming decisions.

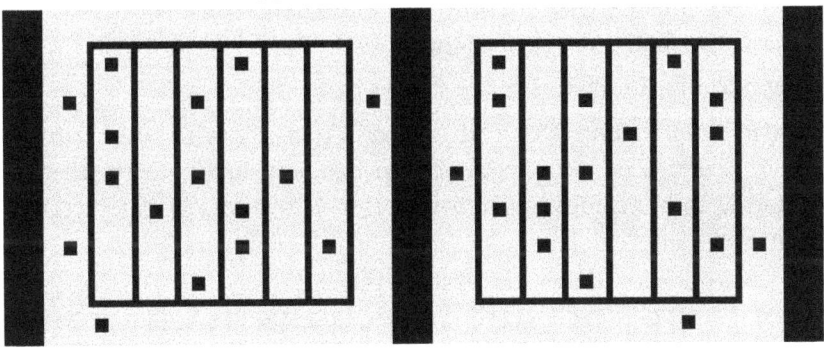

Examining fecal samples using techniques like the McMaster method to help identify the type and quantity of internal worm eggs.

- **Blood Tests:** Blood parameters indicate anemia due to blood-feeding parasites like Haemonchus contortus.
- **Physical Examination:** Veterinarians commonly conduct thorough physical examinations to identify external parasites and assess overall health.
- **Post-Mortem Examination:** In severe illness or death, necropsies (post-mortem examinations) can confirm the presence and extent of parasitic infections.

Common Parasitic Infections in Sheep and Goats

Gastrointestinal Nematodes:

- **Teladorsagia Circumcincta (Stomach Hairworm):** These worms can cause symptoms like diarrhea, weight loss, and anemia in sheep and goats.
- **Trichostrongylus Species:** Small intestinal worms lead to diarrhea, weakness, and reduced feed efficiency.
- **Nematodirus spp. (Thin-Necked Intestinal Worms):** These parasites result in diarrhea and weight loss, particularly in young animals.
- **Coccidiosis (eimeria spp.):** This protozoal infection triggers bloody diarrhea, dehydration, lethargy, and weight loss.
- **Lice and Mites:** External parasites, like the sheep keds (melophagus ovinus) and the scab mite (psoroptes ovis), cause itching, hair loss, skin lesions, and reduced productivity.

Recognition and Diagnosis:

You can do it yourself or call a veterinarian to send fecal samples to detect the presence of worm eggs and spores, conduct a thorough physical examination, and conduct relevant blood tests for appropriate recognition and diagnosis.

Common Parasitic Infections in Chickens

Internal Worms:

- **Ascarids (Roundworms):** Chickens may experience reduced weight gain, decreased egg production, and general weakness.
- **Tape Worms**: These parasites can cause poor weight gain.
- **Cecal Worms (heterakis gallinarum):** Infections with these worms result in poor growth and egg production.
- **Coccidiosis (eimeria spp.):** Chickens may develop bloody diarrhea, dehydration, lethargy, and weight loss.
- **External Parasites (e.g., mites and lice):** Feather loss, skin irritation, and reduced egg laying are common symptoms.

Examining fecal samples for parasite eggs, physical examinations, and a veterinary assessment are necessary for better intervention.

Common Parasitic Infections in Horses

Gastrointestinal Parasites:
- **Small Strongyles (cyathostomins):** These worms can cause colic, weight loss, diarrhea, and lethargy.
- **Large Strongyles (strongylus spp.):** Infections may lead to colic, unthriftiness, and, in severe cases, thromboembolic colic.
- **Ascarids (parascaris equorum):** Young horses with ascarid infections may exhibit coughing, respiratory distress, and intestinal blockages.
- **External Parasites (e.g., ticks and mites):** Restlessness, itching, skin lesions, and reduced overall health are common symptoms.

A vet assessment, fecal sample evaluation, physical examination, and blood tests are mandatory to recognize and diagnose these parasitic infections.

Remember that consulting a veterinarian is essential for accurately recognizing and diagnosing parasitic infections in livestock. Veterinarians have the expertise to interpret diagnostic results, develop treatment plans, and guide parasite control programs, ensuring the animals' health and well-being.

Diagnostic Tools

Diagnosing parasitic infections in animals involves various techniques and tools for the specific parasites and the affected species. Here are some standard diagnostic methods and tools used for identifying parasitic infections in livestock and other animals:

Fecal Egg Counts (FECs)

Fecal Egg Counts (FECs) are commonly used to diagnose parasitic infections, particularly gastrointestinal parasites like nematodes (roundworms). This technique examines animal fecal samples under a microscope to count the number of parasite eggs present. This method is non-invasive, cost-effective, and provides quantitative information about the level of infection.

FECs help veterinarians and livestock owners gauge the severity of parasitic infections, monitor the effectiveness of treatment, and make informed decisions about deworming strategies. However, it's vital to note that FECs may not detect larval or prepatent infections (infections in the early stages). Fresh fecal samples and proper laboratory techniques are essential to obtain accurate results.

Blood Tests

Blood tests are a versatile diagnostic tool for various parasitic infections. They analyze blood samples to detect specific parasite antigens or antibodies produced by the host's immune response. Blood tests can identify many parasites, including blood-borne parasites (e.g., trypanosoma) and some internal parasites.

One of the key advantages of blood tests is that the vet can detect parasitic infections even before clinical signs appear, making them valuable for early intervention. These tests can help with the diagnosis of chronic or latent infections. Specialized laboratory equipment is required to conduct blood tests, and it's not always easy to differentiate between current and past infections in some cases.

Skin Scrapings and Biopsies

Skin scrapings and biopsies are used to diagnose parasitic infections affecting the skin or skin-dwelling parasites like mites and lice. This method collects samples from affected skin areas and is examined under a microscope.

Skin scrapings and biopsies are a direct and accurate method for diagnosing ectoparasitic infestations and skin conditions. This technique requires expertise to collect and interpret samples correctly. It may also miss deep-seated parasites that are not accessible through scraping.

Necropsy

Necropsy, or post-mortem examination, entails the thorough examination of deceased animals to identify internal parasites and understand the impact of these parasites on the animal's health. This method is beneficial for diagnosing internal parasitic infections, including liver flukes.

Necropsy offers a definitive diagnosis of the effects of parasites on internal organs. It is crucial for research, understanding disease dynamics, and monitoring the health of a livestock population. An autopsy requires the sacrifice of the animal and does not apply to living animals.

Serologic Tests

Serologic tests involve the analysis of blood to detect specific antibodies or antigens associated with parasitic infections. These tests diagnose parasitic infections, especially those that cause chronic or latent conditions. They can also indicate exposure to specific parasites. Serologic tests help diagnose chronic or latent infections that other methods may not detect. The interpretation of serologic results can be complex, and the presence of antibodies does not necessarily mean an active infection. Furthermore, the test results may vary depending on the stage of the infection.

PCR (Polymerase Chain Reaction)

Polymerase Chain Reaction (PCR) is a molecular diagnostic technique that amplifies and detects parasite DNA or RNA. It is a highly sensitive and specific method capable of identifying various parasites, including protozoa and certain helminths (worms). PCR offers the advantage of high sensitivity, capable of detecting low-level infections. It is advantageous in cases where other diagnostic methods have failed. However, it requires specialized laboratory equipment and expertise, making it less accessible in some settings.

Ultrasound Imaging

Ultrasound imaging involves using sound waves to visualize internal structures in the body, helping to identify organ-specific parasitic infections. It is frequently used for diagnosing liver flukes and other internal parasites that affect organs. Ultrasound is a non-invasive and real-time imaging method that provides valuable information on organ health and any potential damage or lesions caused by parasites. Carrying out ultrasound imaging requires specialized equipment and training, and it may not detect small or early-stage infections.

Ultrasound is a non-invasive and real-time imaging method that provides valuable information on organ health and any potential damage or lesions caused by parasites.
Langgeng Anggitobumi, CC BY-SA 4.0 <https://creativecommons.org/licenses/by-sa/4.0>, via Wikimedia Commons: https://commons.wikimedia.org/wiki/File:USG_Pada_sapi_Bali.jpg

Skin Swabs and Impressions

Skin swabs and impressions are simple and non-invasive techniques used to collect samples from affected skin areas in animals. These samples are then examined to detect ectoparasites or skin conditions. It's a simple and convenient method that can be performed in the field without requiring specialized equipment. Although skin swabs reveal insights into external parasite infestations, they cannot detect all skin parasites or infections, particularly deep-seated parasites inaccessible through swabbing.

Fecal Floatation

Fecal floatation is a technique used to diagnose animal parasitic infections by examining fecal samples. In this method, the samples are mixed with a flotation solution, causing parasite eggs or cysts to float to the surface, where they can be observed under a microscope. It's a cost-effective method for detecting certain parasites, particularly protozoa and some helminths. It helps identify specific types of parasites and assess the severity of infections.

Histopathology

Histopathology involves the examination of tissue samples under a microscope to identify tissue-dwelling parasites or assess the extent of tissue damage caused by parasitic infections. This method is handy for diagnosing parasites that affect organs or tissues.

Histopathology reveals detailed information about tissue damage and the location of parasites within the affected tissue. It also requires specialized laboratory equipment and expertise for proper sample collection, processing, and interpretation.

Each diagnostic method plays a critical role in identifying parasitic infections in animals. The choice of method depends on the type of parasite, the affected organ system, the animal species, and the case's specific circumstances. Veterinarians and parasitologists use these tools to diagnose infections and develop appropriate treatments accurately.

Handling Samples for Testing

Proper sample collection, storage, and transport are essential to ensure the reliability of the results. Here are guidelines for collecting and handling samples for different diagnostic tests, along with the importance of proper storage and transport:

Fecal Samples (Fecal Egg Counts, Fecal Floatation)
Collection
- Use clean, uncontaminated containers for sample collection.
- Collect fresh fecal samples directly from the rectum or immediately after defecation.
- Ensure that the sample represents the animal's condition and collect enough to perform multiple tests if necessary.

Handling
- Label the sample container with the animal's identification and date of collection.
- Store the sample in a cool, dry place, away from direct sunlight, and seal it to prevent dehydration.
- Avoid contamination from soil or bedding material.

Transport
For fecal egg counts, transport the sample to a laboratory or veterinary clinic as soon as possible. If a delay is expected, store the sample in a refrigerator (4°C), but avoid freezing it, as freezing can damage parasite eggs.

Blood Samples (Blood Tests, Serologic Tests)
Collection
- Use sterile, vacuum-sealed blood collection tubes or syringes.
- Collect blood from a suitable vein, following aseptic techniques.
- Label the sample container with the animal's identification and date of collection.

Handling
- Allow the blood to clot by leaving the sample undisturbed at room temperature for 30-60 minutes.
- Centrifuge the sample to separate serum from clotted blood.
- Transfer the serum to a clean, labeled tube, avoiding contamination.

Storage
- Store the serum sample in a refrigerator (4°C) to prevent degradation.
- Avoid repeated freezing and thawing, which can affect the sample's integrity.

Transport
Transport the serum sample to the laboratory in a leak-proof container, maintaining a cold chain (use ice packs or coolers if necessary) to prevent temperature fluctuations.

Skin Scrapings and Biopsies
Collection
- Collect samples from affected skin areas using a sterile scalpel blade or similar instrument.
- Ensure the samples include the epidermis and any suspected parasites or skin lesions.

Handling
- Place the collected samples in a labeled container, ensuring no cross-contamination.
- Fixative solutions (e.g., 10% formalin) may be used to preserve samples for histopathology.

Storage
Store fixed samples in a cool, dark place, or refrigerate them per specific test requirements.

Transport
Transport the samples to the laboratory in a sealed, leak-proof container, ensuring the sample's integrity is preserved during transport.

Necropsy Samples

Collection
- A qualified veterinarian or pathologist should perform necropsies.
- Collect representative samples of affected tissues or organs, ensuring proper labeling and documentation.

Handling
- Keep the samples separate and well-labeled to avoid cross-contamination.
- Handle samples with care to preserve their integrity.

Storage
- Preserve tissue samples in formalin or other appropriate fixatives.
- Keep samples cool and protected from contamination before they can be transported.

Transport
Transport samples to a diagnostic laboratory, following the laboratory's specific requirements for packaging and shipping.

Proper sample handling, storage, and transport are crucial for obtaining accurate diagnostic results. Failure to adhere to these guidelines leads to sample degradation, contamination, or unreliable test outcomes. Consult with a veterinarian or diagnostic laboratory for specific diagnostic tests and sample type requirements when in doubt.

Chapter 3: Choosing the Right Natural Methods

Using natural parasite control methods for healthy herds can be difficult because they break the accepted norm. Therefore, it can be challenging to receive the correct information and advice. People have become accustomed to using the recommended over-the-counter chemical solutions. There are many benefits to exploring natural pathways to eradicate parasites and deworm your animals. Although the traditional chemical methods are working for now, some parasites are already adapting by becoming resistant to commonly used commercial dewormers. Therefore, the need for finding alternates as the traditionally used chemical options become ineffective is increasing exponentially. Moreover, the environmental impact of pharmaceutical production and transportation is antithetical to running a sustainable farm.

The wide range of animals and species-specific parasite relationships makes it difficult to apply general care. Equipping yourself with the appropriate knowledge unlocks a new world of sustainable treatment of your livestock. Grabbing hold of natural remedies for parasites opens the door to an ancient tradition that has built up tried and tested systems for centuries. Using plants and herbs for parasite control is a practice found around the world. Ancient Nordic cultures brewed carefully mixed plant medicines for their livestock, and people still report the effectiveness of these methods in these regions.

Using plants and herbs for parasite control is a practice found around the world.
https://www.pickpik.com/herbs-french-bouquet-gourmet-cuisine-rosemary-38483

As the movement for more sustainable practices grows, there will be a higher demand for remedies coming from local geographical areas. The rising environmental consciousness among many farmers has generated the development of parasite control using ingredients from their local region. Therefore, the research on natural deworming and parasite control is steadily growing. By accessing what is known and refining it to your needs, the animals you are raising, and the environmental factors unique to your ecosystem, you can extract all the benefits from natural remedies while minimizing their downsides.

Natural Deworming and Parasite Control

Natural deworming and parasite control is the use of herbal and plant solutions to treat your livestock as opposed to pharmaceutical medications that can be bought over the counter or acquired from a veterinarian. Someone motivated to use natural deworming options could be pursuing this way for numerous reasons, including parasite resistance being detected and wanting to be more eco-friendly in their farming approaches. Natural parasite management is linked to an entire philosophy that emphasizes working within the conditions of local habitats to achieve your desired farming outcomes. Following the guidelines of natural deworming is a regenerative practice that promotes mutually beneficial bonds between nature, yourself, and your livestock.

When people think of breeding organic livestock, the first thing that often comes to mind is the use of antibiotics. The development of antibiotic-resistant bugs caused many farmers to opt for the organic route. However, with the focus on the negative results of overusing antibiotics, natural antiparasitic medications (called *anthelmintics*) were overlooked. To be truly organic and reconnect with nature and healing means a complete overhaul of how you treat and maintain your livestock must be conducted. Natural deworming is not as simple as popping a few tablets in your feed. You must consider what you plant in your pasture and how your food chain connects at the micro and macro levels. By being observant and aware, you better understand the complex natural networks that result in healthy livestock.

When you look at your farm as an interdependent ecosystem and why organisms interact in mutually beneficial or hostile ways, you start to see the bigger picture of parasite management. Chemical dewormers are too narrow in their focus because they zoom into the parasite as a problem for the animal instead of viewing the broader scope of the environment. The animals you choose to keep, the pastures you plant, how you manage them, and the cleanliness of your farm all play a role in creating a situation where parasites thrive or where they are kept down to manageable rates.

When you use natural antiparasitic methods, you have to be more present in the existence of your livestock. Animals affected by parasites will behave differently and show signs of needing deworming. You also need to pick up if your pasture environment is geared toward being a breeding ground for parasites. Being able to observe these nuances comes with careful education and practice. Adopting natural ways to eliminate parasites requires a lifestyle shift because your farm or homestead needs to be restructured following the holistic approach nature demands.

Parasite management is more than a visit to the vet. You must understand both the life cycles of the animals you farm, how different species relate to one another, and the life cycles and functioning of parasitic organisms. A parasite may live inside an animal for some of its life and get transferred to the next stage through fecal matter. Once you integrate natural parasite control into your routines, you'll notice that the methods encourage biodiversity, and sourcing what you use locally positively contributes to the regeneration of the environment. From the people who consume meat to the plants, animals, and farmers – all benefit from your commitment to transitioning to using homegrown natural remedies and techniques to eliminate parasites.

Why Choose Natural Methods over Artificial Chemicals

There are pros and cons to using natural antiparasitic techniques and artificial chemicals. When you weigh up both of these methods, the natural way comes out far ahead of anything chemical on the market for many reasons. When you examine which is better, the naturalistic approach or chemical treatment, you must first explore the relationship between animals and parasites. The first point to note is realizing that every farm has parasites on it, so there is no running away from the fact that your livestock will be affected at some stage. The key to parasite management is minimizing the impact on your herd by reducing the number of affected animals and treating the ones that have already been infected.

The main attraction to using chemical antiparasitic medicine is the convenience. Once you know what parasite affects your herd, you go to the supplier and buy what you need. If a vet visits your farm, they tell you which product you need and how to use it. However, the convenience of using these pharmaceutical chemicals has caused many large-scale commercial farms to use chemicals over natural methods. This widespread use has made some parasite species immune to commonly used medicines. Some dewormers will have to be phased out because of the rising resistance to the medication. When parasites are expelled using chemical means, some of them survive. With their stronger genetics, the surviving parasites reproduce, birthing a new generation of pests immune to the chemicals that killed off the rest of their kind. Natural methods do not have the same side effects.

Other than developing super parasites resistant to dewormers, natural antiparasitic measures require you to adapt to more environmentally friendly farming methods. Since you are no longer using chemical dewormers, your animals will be unable to survive in the same conditions they once did. On inorganic farms, animals live on top of one another and are simply injected with all kinds of artificial medicine to avoid the spread of diseases and parasites that thrive in these cramp conditions. One tenet of natural parasite management is providing enough space and appropriate living conditions that are not a breeding ground for the killer critters. When you use pasture grounds and allow animals to roam in free space, they have a better quality of life. Therefore, natural parasite control is better for your livestock based on their comfort.

Dewormers can adversely impact the ecosystem, inadvertently killing creatures that work for the benefit of your farm. Chemical antiparasitic medicine may get rid of the parasites, but some of the artificial compounds are released in your livestock excrement. This penetrates the soil and kills organisms central to the environment's biodiversity and function. For example, dung beetles and earthworms could be mistakenly killed by dewormers, which will reduce the quality of your soil. The knock-on effect is that low-quality soil will then affect the crops you can grow, as well as your grazing pastures for your livestock, which means your animals will get low-quality feed. This impacts their muscle development and the birthrate of breeding animals. Farmers in Scotland started experiencing problems with the sheep herds due to the impact of dewormers on the soil. It was suggested that they reduce their use of dewormers or cease altogether. Some farmers in the region took a more targeted approach by treating only affected animals. Natural methods offer the perfect solution to the soil degradation problem.

Considerations to Make before Pursuing Natural Parasite Management

Transitioning from chemical antiparasitic solutions to natural ones can take a lot, especially at first. Each farm and set of circumstances is different, and much like nature adjusts, your approach to parasite control will shift. The livestock species, your region, and the parasites your animals are most affected by will all play a role in what techniques you implement. Jumping in head first can be tempting, especially if you are excited to explore this fulfilling journey. However, for your safety and the well-being of your animals, you need to slow down and conduct a thorough analysis of your farm. Making a few basic considerations that will affect how you approach natural deworming will increase your chances of success and avoid devasting failures that can cost you a large chunk of your herd.

Parasite Species

Parasites take many shapes and forms. For example, parasites can live inside an animal's gut, like many worm species, or on the skin's surface, like fleas and ticks. Each body part and organ of your livestock can be affected by parasitic species. Therefore, before you plan to reduce parasites, you must know which species you are dealing with and what your livestock is particularly susceptible to. Some of the parasite groups common to the geographical location that you should research are endoparasites that live inside the host, ectoparasites that live in their bodies, hemoparasites that are in the blood, or even protozoa which are unicellular microscopic parasites.

Livestock Species

The types of livestock on your farm will shape your choices for parasite control. For example, sheep and cows do not share the same parasites, but goats and sheep have some common parasites. Therefore, you must consider which animals you keep and how you raise them to prevent the spread of different parasite species. For example, grazing animals will often get reinfected with parasites because they are exposed to them more when they go out into a pasture. Keeping your grass long is one way to reduce infection rates with grazing livestock. Many parasite species only exist on the first couple of inches of grass. So, if your grass is long, your livestock will mostly eat the upper levels where there are fewer parasitic species.

Environmental Conditions

The habitat conditions your animals live in will determine which kinds of parasites develop. For example, after floods, there is usually an upsurge in parasites. Warm, wet areas will have particular parasite species that differ from desert climates. Furthermore, the hygiene conditions on your farm will also impact which parasites are around. For example, some farms keep chickens in battery cages, so they are constantly surrounded by fecal matter, which means they easily pick up parasites.

Diseases and Illnesses

Certain of your livestock will be more susceptible to dying from parasite infections due to pre-existing health issues. One of the natural ways to combat parasites is by selectively breeding your livestock. If a parasite infection spreads through your herd, separate the ones that seem the least affected and breed them. Their genetics will act as the first barrier against the negative impact of parasites. Younger animals are also usually more likely to die from parasites. Therefore, you must be mindful of the life stages of your livestock, common parasites that affect them, and what illnesses leave your herd immunocompromised. For example, you may want to keep animals recovering from certain illnesses separated from the herd to avoid infection while they are in a vulnerable state.

Proactivity and Communication with Medical Professionals

You need to consult medical professionals for effective parasite treatment and eradication, whether chemical or natural. Being proactive in contacting the vet requires you to be aware of some of the telltale signs that you have an infected herd. Different animals express illness differently when parasites infect them. Some key tips to remember are to check their energy levels, check their excrement for eggs, and examine their skin to see if there are wounds or rashes that parasitic infections can cause. If you identify one animal with worms or other parasites, it is likely that more animals in your herd are also affected. Separating affected animals is one of the first steps you should take before calling a veterinarian.

Sometimes, it is almost impossible to tell if an animal has parasites. For example, cattle with small intestinal worms and brown stomach worms show no signs of infection until the worm load is so high that it causes sudden death. When you have mysterious deaths on your farm and your animals seem otherwise healthy, it may indicate that you need to take some steps for parasite control. Calling the vet to look at the animal and the environment will help you know which parasite you are dealing with. The tests that vets run are more accurate than your calculated guesses. Even if you opt not to use the chemical pharmaceuticals vets recommend, the information they give you will guide you on what natural measures to take. If you openly communicate with your vet that you are using natural antiparasitic techniques, they can also advise you on what natural measure you should take.

An informed opinion is essential for making choices that will benefit you and your animals. Building a relationship with your vet allows you to reach out when you have concerns, and your proactivity includes your vet in advising you on your goals for animal health. Monitoring your animals closely allows you to notice anomalies earlier to catch any problem and have a better chance of catching it. Furthermore, since prevention is better than a cure, your vet will tell you what to do so your livestock is not susceptible to diseases or parasites.

You should schedule regular health checks for your herd by your veterinarian and not only rely on them for emergencies. Vets can advise you on the environmental conditions on your farm, the nutrition of your animals, as well as what medical interventions are required for your herd. Animals on a farm require constant care, and if you want to maintain or grow the numbers of your herd, it is crucial to know what is going on with their health. Therefore, proactivity in contacting your vet for regular checkups will benefit you in the long run.

A strong relationship with your vet as a team member enhances your farm in many ways. Think about going to the dentist. If you get a check-up every six months, you can address the small problems that arise along the way, and the dentist will tell you where you are going wrong in caring for your teeth. However, going to the dentist only when you have a toothache will cost you a lot more. Similarly, calling your vet occasionally will result in them participating in disaster management instead of playing their role to elevate your farm. Just like cleaning, feeding, and rotating animals on your pastures are part of your maintenance routine, veterinarian visits should be included in that list for an elite farm environment to be created that can facilitate natural antiparasitic protocols.

Creating a Deworming Plan

Having a planned schedule, as opposed to leaving deworming to the times when your livestock is already infected, is part of farming best practices. Your deworming plan, especially if you do it naturally, is about keeping a healthy, productive farm running. For example, if you have a variety of animals on your farm, you can allow your sheep and horses to graze together, but you should prevent your sheep and cattle from mixing due to cow parasites having deadly impacts on sheep herds. In this way, you can prevent using excessive deworming products, even if they are plant-based and herbal.

Goals

Your goals, as they relate to parasites, should be realistic. For example, completely eradicating all worms, ticks, or flies is impossible to achieve. Set measurable goals that are achievable and realistic. Making year-over-year comparisons will help you craft the goals to achieve with your deworming strategy. So, if you lost a certain amount of livestock to parasites the previous year, you can set a goal to lose fewer animals this year. Other goals can be related to how you organize your farm or, if you've been using chemical deworming methods, how to transition to natural options.

Needs

Every farm has individual needs depending on where it is situated and what combination of animals are being kept. For a plan to be effective, list every animal you have and the parasites that affect those animals in your location. Next, you must survey your land to see how you can best accommodate all your animals and which grazing patterns you can use to minimize their risk of getting parasites. Lastly, you must find which plants, herbs, and foods help the animals expel parasites in the numerous forms they take. Finally, you must set up emergency protocols for quarantining your livestock when necessary.

Time Management

Your daily schedule must be set by the hour so that you can do all the cleaning, culling, breeding, isolating, and feeding activities needed to keep parasites at a minimum in your herd. Remember that you must always check your animals and their excrement for signs of parasites, which takes significant time out of your day. To prevent you from feeling overwhelmed, managing your time well and scheduling every activity you must complete for the day, month, week, and season is best.

Environmental Factors

Different weather patterns and seasons bring parasites along with them. Furthermore, the habitat where you raise your animals also has an ecosystem where parasites slip neatly into the food chain. You have to draft measures to accommodate these factors. For example, after floods, stable flies are common. Therefore, you must have a preparation plan that expects this increase after the rainy season. Your animals live in tune with nature, which does not always mean positive outcomes. Nature often means suffering for animals in many cases. Therefore, you must understand the natural world around your farm to respond to it efficiently and promptly.

Chapter 4: Grazing the Parasites Away – Pasture Management

In sustainable livestock farming, one of the biggest challenges is dealing with parasites lurking in your pastures. To tackle this issue, you need a well-thought-out pasture rotation system, regardless of whether you're raising one type of animal, a mix of species, or using a specific grazing method. This system is all about controlling how long animals graze in a given area and managing their grazing habits.

A remarkable and intricate countdown begins once the parasite eggs are expelled from the host and deposited in feces. The eggs hatch within one to 14 days, usually around five days. After hatching, some parasite larvae can hang around for up to 40 days if the conditions are right – think moist and warm ground. But after about 55 days, most of these larvae will be gone unless they find a new host before that.

Knowing how parasites work can help you plan your grazing strategy to prevent your animals from getting reinfected. For example, in just four days, the notorious barber pole worm, which can harm goats and sheep, can hatch and pose a threat. Since these larvae only survive on the ground for about four weeks, it's smart to move your goats or sheep to fresh grazing areas every four days or even more often when parasite levels are high. After this intense grazing period, let those sections of pasture rest from goat or sheep grazing for at least 60 days to ensure all parasite larvae are gone. You can do this by dividing your pastures using temporary or electric fencing.

Pasture management is a great way to deal with parasites. As you learn more about pasture management, you'll discover different strategies and techniques that protect your animals from parasites and help you create a sustainable and healthy grazing routine.

Grazing as a Natural Approach to Parasite Management

Grazing in Parasite Control is a natural approach to managing and minimizing parasitic infections in livestock. It involves strategically controlling the movement and grazing patterns of animals on pasturelands to disrupt the life cycles of common livestock parasites, including nematodes (worms), coccidia, and other harmful microorganisms. Some of the different approaches you can take are set out here:

1. Strategic Grazing

Grazing in parasite control encompasses carefully planned rotation and pasture allocation. The key idea is to avoid overgrazing in one area and to give your pastures adequate resting periods, which is essential in breaking the parasite life cycle.

The key idea is to avoid overgrazing in one area and to give your pastures adequate resting periods, which is essential in breaking the parasite life cycle.

https://pixabay.com/photos/cow-grassland-grazing-nomad-7200409/

2. Rotation and Rest

This approach involves rotating animals through different pastures and allowing previously grazed areas to rest. During the resting period, the absence of host animals interrupts the completion of parasite life cycles, resulting in a decline in the overall parasite burden.

3. Controlled Grazing Intensity

Grazing intensity and duration are managed to ensure that animals do not have prolonged exposure to parasite-infested pastures. By reducing the time animals spend in a particular area and then moving them to fresh pastures, the risk of reinfection by parasites significantly decreases.

Understanding the Life Cycles of Parasites

To manage and control parasites in livestock effectively, it's essential to grasp the intricacies of their life cycles. Different parasites, such as nematodes and coccidia, have specific life stages and dependencies on environmental conditions. Understanding these life cycles is key to developing successful parasite control strategies. For instance, the barber pole worm (haemonchus contortus) affects small ruminants like goats and sheep. Its life cycle involves egg shedding in the host, hatching into larvae that develop in pasture, and reinfection of the host through ingesting infective larvae. Recognizing this cycle is vital to addressing barber pole worm infestations.

Grazing management is an effective tool for interrupting the life cycles of common livestock parasites. By putting well-planned rotational grazing and pasture rest periods in place, you can break the chain of infection. Through rotation, you move livestock to different pastures. During the resting phase, the absence of host animals prevents the completion of parasite life cycles. For example, the larvae shed in feces require host contact to develop into adult worms. Consider a rotation cycle that moves goats to a new pasture every 35 days. This allows the previously grazed pastures to rest for at least 60 days. During this period, the majority of parasite larvae will perish.

Controlling grazing intensity and duration is essential. Limiting animals' time on a particular pasture and then relocating them to fresh ones reduces the risk of reinfection. Rapid movement to new pastures prevents animals from consuming infective larvae lingering in their previous grazing areas. If you have a high-risk parasite like the barber pole worm, you may have to move goats every four days during peak infestation periods, preventing larvae from becoming infective.

Grazing Strategies

1. Custom Grazing Habits

When designing a rotation system, pay attention to your livestock's specific habits and needs. Different species have unique preferences, and tailoring your grazing practices to their behavior can be highly effective. For instance, if you consider goats and cattle, goats are browsers and prefer various forage types, while cattle are primarily grazers and focus on grass. Consider this when planning which pastures to allocate for each species to maximize their nutrition intake and minimize parasite exposure.

2. Pasture Orientation

The orientation of pastures also plays a role in parasite control. You will notice that the larvae dry out faster on your south-facing slopes than in other pastures, particularly during spring. The combination of sun and slope reduces moisture and larval survival. If you have a south-facing pasture available, reserve it for use in spring when moisture levels in the soil and grass are higher. This combination promotes rapid larval desiccation, decreasing the risk of livestock ingesting infective larvae.

3. Understanding Larvae Movement

Recognizing how larvae behave in different conditions is crucial. Wet grass encourages larvae to move away from feces, while dry conditions keep them closer to the ground. Understanding these patterns helps you to determine your grazing strategies. For instance, in wet conditions, larvae may be found up to a foot away from feces, making it important to move livestock frequently to prevent ingestion. In dry conditions, larvae stay closer to the ground, reducing the risk of consumption.

4. Impact of Weather

Weather conditions can influence parasite larvae behavior. Larvae tend to climb higher on plants during overcast periods, like rainy days or early mornings and late afternoons, to avoid bright light. Being aware of these conditions helps in managing grazing schedules. Avoid grazing livestock during early mornings or late afternoons on overcast days or right after rain, as this is when larvae are more likely to be higher on plants. Instead, plan grazing sessions for drier and brighter periods.

5. Electric Fencing

Implementing one or two strands of electric fencing is a practical way to divide pastures into sections. This approach enables you to move livestock frequently, control their access, and prevent overgrazing. Create temporary sections within your pasture using electric fencing. After a certain period, move livestock to the next section, allowing the previous one to rest. This not only controls parasite infection but also optimizes forage utilization.

Seasonal Pasture Management

Seasonal pasture management involves tailoring your grazing practices to leverage the changing conditions and particular requirements of each season to maximize livestock nutrition and effectively disrupt parasite life cycles. In this approach, the seasonal rotation of livestock is guided by the conditions of the time of year. Whether you are taking advantage of the warming spring sun to desiccate parasite larvae or utilizing lush summer pastures to limit parasite exposure, these strategies aim to promote both the well-being of your livestock and the control of parasitic burdens.

1. Spring Pasture

During spring, focus on pastures that face south and have a sloping terrain. These features maximize sun exposure and efficient drying of the pasture. As a result, it reduces moisture levels, which is critical for parasite larvae survival. For example, a south-facing pasture with good sun exposure helps dry feces and the surrounding soil, decreasing the likelihood of larvae survival. Grazing in such pastures during this season can help control parasite infestation.

2. Summer Pasture

In hot summer months, lowland grazing areas with lush vegetation are advantageous. The heat forces parasite larvae to stay closer to the ground and moist manure, making them less likely to be consumed by grazing animals. Summer pastures in lowland areas with ample grass growth are ideal. As the sun intensifies, larvae remain close to the ground, reducing the risk of livestock ingesting them.

3. Fall Pasture

In the fall, diversify your livestock's diet by offering nuts, fruits, and leaves. However, monitor moisture levels and temperatures, as some parasites can survive under leaf cover. Goats, sheep, and other livestock species can benefit from the nutritional variety provided by these additions to their diet. Be cautious and monitor weather conditions to ensure that parasite survival is minimized.

4. Winter Pasture

Winter pastures in regions with USDA Hardiness Zones of 6 or higher can include taller, "stocked" forage. The cold winter conditions naturally reduce parasite populations. Utilize winter pastures with taller forage growth. The combination of cold temperatures in Zone 6 or higher and the absence of hosts on the pasture helps decrease parasite burdens during the winter months.

Other Grazing Methods

1. Multi-Species Grazing

Multi-species grazing, especially during heavy growth periods, can be a powerful tool to prevent overgrazing and reinfection. When different livestock species graze together, they have varying preferences for forage types and different grazing habits, reducing the pressure on specific plants and mitigating the risk of parasite transmission. For example, combining cattle and goats in a multi-species grazing system can be very effective. Cattle primarily graze grasses, while goats are browsers and prefer shrubs and forbs. This diversity in grazing behavior can help minimize the impact on any single forage type.

2. Mixed Livestock Grazing

Mixed livestock grazing involves combining different species on the same pasture, such as cattle and goats or horses and sheep. However, it's crucial to avoid grazing sheep and goats in succession since they share many parasite species, and consecutive grazing can lead to increased parasite burdens. Grazing cattle and horses together is an option, as their dietary preferences and digestive systems complement each other. Cattle primarily graze grasses, while horses are more selective, often avoiding certain grass species. This combination can lead to more efficient forage utilization.

3. Timing Is Crucial

Timing is essential when moving livestock back onto pastures, particularly during wet conditions. Rushing livestock back before 60 days have passed can lead to increased parasite levels. Instead, schedule grazing during dry, bright daylight hours to minimize exposure to infective larvae.

During wet seasons or in rainy areas, ensure that pastures have adequate time to rest between grazing sessions. Grazing during dry, sunny hours when the grass and ground are less likely to harbor infective larvae is prudent.

Grazing during dry, sunny hours when the grass and ground are less likely to harbor infective larvae is prudent.
https://www.rawpixel.com/image/3260248/free-photo-image-livestock-farm-calf-yard

4. Grazing Height

Maintaining the ideal grass height is crucial for both optimal forage quality and effective parasite control. Aim to keep the grass between 6 to 8 inches in height. Grass that is too short makes it easier for livestock to ingest infective larvae, while excessively tall grass limits eggs and larvae from being dried out by sunlight. Regularly measure the grass height in your pastures to ensure it falls within the 6 to 8-inch range. This promotes better forage utilization and limits parasite exposure by keeping the grass at an ideal height.

Beneficial Soil Organisms

1. Predatory Nematodes

Predatory nematodes are microscopic worms that play a crucial role in reducing parasite populations. These nematodes are natural predators of parasite eggs and larvae, effectively controlling their numbers. For instance, consider Steinernema carpocapsae, a predatory nematode species highly effective against various parasite larvae, including houseflies, stable flies, and flea larvae. You can significantly reduce parasite levels by releasing these nematodes into your pastures.

2. Beneficial Beetles

Beneficial beetles, including ground beetles, rove beetles, and dung beetles, are essential components of a healthy pasture ecosystem. They help control parasite larvae and contribute to improved soil and plant health. Dung beetles are particularly valuable for parasite control. There are three types of dung beetles: rollers/tumblers, tunnelers, and dwellers. Rollers and tunnelers bury manure beneath the ground, effectively trapping parasite eggs and larvae. This not only prevents hatching but also improves soil and plant health.

Dung beetles, especially the tumblers and tunnelers, significantly impact parasite control. They work diligently to bury livestock manure, ensuring parasite eggs and larvae are too deep within the soil to reach the surface. As dung beetles bury manure, they also bury any parasite eggs and larvae present in the feces. This disrupts the life cycle of parasites and promotes the drying and deactivation of eggs and larvae, reducing their infectivity.

3. Scattered Manure

Due to dung beetle activity or other factors, scattered manure dries more quickly than clumped or concentrated manure. The quicker drying process shortens the opportunity for parasite larvae to remain moist and infectious. In pastures with active dung beetle populations or efficient manure scattering, you'll notice that manure is dispersed across the pasture. This scattered manure dries rapidly, making it an inhospitable environment for parasite larvae.

Environmental Control

You can implement measures to control deer access to pastures, especially those designated as harmful to goats, sheep, llamas, alpacas, and calves. Deer can serve as hosts for certain parasites and add to the contamination of pastures. Use top-strand electric fencing or guardian dogs to deter deer from grazing in pastures. Limiting deer access reduces the risk of parasite transmission to your animals.

You can also use fowl, like ducks, chickens, and other poultry, to manage gastropods, which are intermediate hosts for various parasites. Fowl can help control snail and slug populations, reducing the risk of livestock ingesting these parasites. You can allow ducks and chickens to forage in pastures prone to gastropod infestations. Their natural behavior of pecking at snails and slugs limits the numbers of these intermediate hosts, contributing to reduced parasite burdens in your livestock.

Environmental Maintenance

Effective environmental maintenance is key to managing parasite infections in livestock and promoting overall farm health. Consider the following steps to create a conducive environment that discourages parasites and supports beneficial organisms:

1. Reduce Close-Cut Mowing and Excessive Sun Exposure

Moisture-dependent organisms, like predatory nematodes and dung beetles, require some shade to thrive. Avoid close-cut mowing and extended exposure to constant sunlight, which creates ideal conditions for these beneficial organisms. A balance between sun and shade allows for the coexistence of these organisms and reduces the potential for parasite contamination. So, maintain a mixture of shaded and sunny areas within your pastures to support the activity of moisture-dependent organisms.

2. Avoid Chemical Usage

Refrain from using chemical dewormers, herbicides, pesticides, and other chemical compounds in your pastures. Chemical substances can disrupt the life cycles of beneficial organisms, reducing their effectiveness in parasite control. Choose natural and organic alternatives to minimize the harm to these beneficial creatures. This practice helps to preserve the populations of predatory nematodes, dung beetles, and other beneficial soil organisms.

3. Promote Clean Water Systems

Keep control of clean and efficient water systems for your livestock by eliminating standing water puddles. Stagnant water is a breeding ground for parasites and protozoa, increasing the risk of infection. Encourage livestock to drink from clean, regularly maintained water tanks. When implementing rotational grazing, use movable tanks with floats that can be easily moved from section to section. Regular cleaning and filling of these tanks ensure that livestock have access to clean water, reducing the risk of waterborne parasite infections.

4. Keep High-Traffic Areas Clean

Regularly clean and maintain paddocks, stalls, feeding areas, and other high-traffic areas. Damp debris, like loose hay and leaves, can protect the parasites and increase the likelihood of livestock ingesting them. Keeping these areas clean and dry minimizes the risk of parasite contamination.

After cleaning, apply drying agents such as barn lime, sulfur powder, wood ash, or diatomaceous earth to the surfaces. These agents help dry out any remaining parasite eggs and larvae, further reducing the risk of infection.

Including these pasture management and environmental control practices will help reduce the risk of parasite infection and ensure a healthier and more productive herd.

Limitations of Grazing-Based Parasite Control

1. Climate Variability

Climate variability can significantly influence the effectiveness of grazing-based parasite control. Different weather conditions, such as prolonged periods of rain or drought, can impact parasite survival and transmission. It's essential to consider these challenges and adapt your grazing strategies accordingly.

- **Rain and Moisture**

    Excessive rainfall and high humidity can create ideal conditions for parasite survival, making it challenging to control parasite populations. In such cases, adjusting your grazing rotations, reducing stocking rates, or implementing other parasite management techniques may be necessary.

- **Drought and Limited Forage**

    Conversely, drought conditions can limit forage availability, which might lead to overgrazing and increased parasite exposure. During droughts, you may need to provide supplemental feed and water to keep your livestock healthy and reduce the risk of parasite infection.

2. Considerations for Specific Livestock Species

Different livestock species have varying susceptibility to parasites. Understanding your livestock's specific needs and vulnerabilities is crucial for effective parasite control. For example, goats and sheep are more susceptible to barber pole worms, while cattle face different challenges.

- **Species-Specific Management**

    Tailor your grazing management to what your livestock needs. Consider their natural behaviors, dietary preferences, and parasite susceptibility. Implement rotational grazing that aligns with the specific requirements of each species.

- **Selective Grazing**

    Some livestock, like goats, are known for their selective grazing habits. Utilize this behavior to your advantage by planting parasite-repelling plants in their pastures, encouraging them to self-medicate.

Extra Tips for Good Pasture Management Practices

1. Soil Health and Fertility

Soil health is fundamental to pasture management as it directly impacts forage quality, livestock nutrition, and overall farm productivity. Implement practices to improve and maintain soil fertility.

- **Soil Testing**: Conduct regular soil tests to assess nutrient levels, pH, and other factors. Soil testing provides essential data to guide your fertilization and liming strategies.
- **Fertilization**: Based on soil test results, apply appropriate organic or mineral fertilizers to correct nutrient imbalances and enhance soil fertility. Well-fertilized pastures support lush, nutritious forage.
- **Liming**: Adjust soil pH using agricultural lime to create a more favorable environment for nutrient uptake by forage plants.
- **Organic Matter**: Increase organic matter in the soil through composting, cover cropping, and rotational grazing practices. Higher organic matter content improves water retention, aeration, and nutrient cycling.

2. Monitoring and Record-Keeping

Effective pasture management requires regular monitoring and diligent record-keeping to make informed decisions and track changes.

- **Pasture Walks**: Schedule frequent pasture walks to assess forage growth, plant health, and any signs of overgrazing or parasite issues.
- **Record Livestock Movements**: Maintain records of livestock rotations, including dates and pastures used. This information helps prevent overgrazing and optimize parasite control.

- **Weather Observations**: Keep a weather journal to record conditions such as rainfall, temperature, and humidity. These factors influence grazing strategies and parasite activity.
- **Growth Data**: Measure and record forage height and density to determine when to rotate livestock and prevent overgrazing.

3. **Forage Selection and Pasture Renovation**

The selection of appropriate forage species and regular pasture renovation are key parts of maintaining productive and healthy pastures.

- **Forage Diversity**: Diversify your pasture with a mix of forage species that suit your region and livestock. Choose varieties that provide a balance of nutrition and palatability.
- **Renovation Practices**: Implement overseeding and reseeding as part of pasture renovation. These practices help introduce new, productive forage varieties and improve overall pasture quality.
- **Grazing Heights**: Set grazing heights to maintain healthy pasture plants. Avoid overgrazing, which weakens plants, and undergrazing, which allows weeds to proliferate.
- **Rotational Grazing**: Continuously rotate livestock through pastures to prevent overgrazing and allow forage to recover.

Pasture management is a cornerstone of sustainable livestock production, offering numerous benefits. It leads to improved animal health and performance through balanced nutrition and reduced parasite loads. It minimizes environmental impacts, including soil erosion and nutrient runoff, while enhancing forage production and soil health. Pasture management allows for reduced reliance on chemical dewormers and other synthetic interventions, contributing to a balanced ecosystem that supports biodiversity and beneficial organisms.

Chapter 5: Nutrition and Immunity

Nutrition and immunity in livestock are intricately connected. The quality and balance of the animals' diet are pivotal in bolstering their immune system. Essential nutrients like proteins, vitamins, and minerals are necessary for a healthy immune function. Well-nourished livestock are better equipped to resist infections, recover from illnesses, and maintain overall health than animals that are kept nutrient-deficient.

Inadequate nutrition weakens the immune response, making animals more vulnerable to diseases. Therefore, ensuring a proper and balanced diet is necessary to optimize the immune defense systems in livestock. Feeding the proper nutrients through a well-balanced diet prevents diseases, enabling them to ward off infections and remain resilient to various health challenges.

Nutrition, the immune system, and parasite control influence each other. Here's the critical link between these three factors for a clear perspective.

Nutrition and the Immune System

Proper nutrition supports the immune system's functionality. A well-nourished animal is better equipped to have an adequate immune response when exposed to pathogens, including parasites. For example, protein is essential for producing antibodies and immune cells, while vitamins and minerals play critical roles in various immune functions.

Proper nutrition supports the immune system's functionality.
https://pixabay.com/photos/goats-lambs-animal-goats-lambs-5110369/

Nutrition and Parasite Control

Although every nutrient has a role in strengthening the immune system and protecting livestock from diseases, certain nutrients like zinc and selenium directly improve the animal's resistance to parasites. The natural defense mechanism of animals weakens when some nutrients are not included in a balanced diet, increasing the chances of developing parasitic infections.

Immune System and Parasite Control

As you already know, the immune system is the first line of defense against parasites or other infectious diseases. When an animal is exposed to parasites, a properly functioning immune system can identify, attack, and control the parasites' populations within the host's body. This is particularly necessary in preventing parasitic infections from becoming severe or causing long-term health issues.

Striking a balance between adequate nutrition, caring for the immune system, and parasite control is imperative to maintain the health and productivity of the livestock. Proper nutrition supports a robust immune response, which, in turn, helps in preventing and controlling parasitic infections. Farmers and livestock managers must feed well-balanced diets, manage pasture rotation, and use appropriate veterinary treatments to ensure the health and well-being of their animals while minimizing the risk of parasitic infections.

Why Feed a Balanced Diet

Providing a balanced diet to livestock is necessary for several reasons mentioned here:

Nutrient Support

A balanced diet includes essential nutrients like proteins, carbohydrates, fats, vitamins, and minerals in appropriate proportions. Besides maintaining the functioning of the immune system and developing adequate immune responses, nutrients are required in thousands of other metabolic processes. Inadequate or imbalanced nutrition only weakens the immune system and can trigger certain medical conditions, ultimately making animals more susceptible to infections.

Energy and Maintenance

Livestock require a sufficient energy supply to keep their basic bodily functions healthy and support growth, reproduction, and milk production. A well-balanced diet ensures that animals have the energy to carry out these functions efficiently. When animals are malnourished, their energy levels drop, making them more vulnerable to diseases and less capable of maintaining their overall health.

Disease Resistance

The immune system relies on nutrients like vitamins (e.g., vitamin C, vitamin D), minerals (e.g., zinc, selenium), and proteins to function effectively. These nutrients are involved in antibody production, white blood cell activity, and other immune responses. A balanced diet ensures livestock receive the necessary building blocks to mount a robust defense against pathogens. Conversely, poor nutrition leads to immunosuppression, reducing the animal's ability to resist diseases.

The immune system relies on nutrients like vitamins (e.g., vitamin C, vitamin D), minerals (e.g., zinc, selenium), and proteins to function effectively.
https://unsplash.com/photos/person-holding-orange-plastic-bottle-eUGppEZgkAM

Growth and Reproduction

Healthy animals are more likely to reach their growth potential and produce offspring successfully. Malnourished animals may experience growth stunts and reproductive problems, making them more susceptible to diseases due to weakened physiological states.

Maintenance of Body Condition

Adequate muscle mass is essential for locomotion and the ability to escape from disease vectors and predators. Furthermore, maintaining an appropriate body condition ensures that animals have adequate fat reserves, which can be crucial during stress, such as cold weather or disease outbreaks. The meat quality also improves when adequate nutrients are fed to the animal.

Preventing Metabolic Disorders

Certain diseases and conditions in livestock are associated with imbalances in the diet. For example, overfeeding grains to cattle can lead to conditions like acidosis, while inadequate calcium and phosphorus levels can result in disorders like milk fever in dairy cows. A balanced diet prevents metabolic disorders like these, which can harm animal health.

Resistance to Stress

Proper nutrition enhances an animal's ability to cope with various forms of discomfort, like stress and agitation, which is evident during transportation, changes in environmental conditions, and the introduction of new animals to the herd. When livestock are well-nourished, they are more resilient to stressors, which can otherwise suppress the immune system and make them more susceptible to diseases.

Lowering the Risk of Zoonotic Diseases

Balanced nutrition can also affect food safety and human health. Some diseases in livestock are zoonotic, meaning they can be transmitted to humans. Maintaining disease resistance in livestock through proper nutrition reduces the risk of zoonotic diseases, benefiting nearby animal and human populations.

Dietary Components for a Robust Immune System

The amount and type of feed change depending on the livestock type. Cattle require a mix of hay, grass, grains, silage, and legumes, whereas goats are mostly kept on dry or green forage with added protein and mineral concentrates. Here are the essential components dietary must include for adequate health and well-being.

- **Protein:** This is essential for producing antibodies, enzymes, and immune cells central to the immune response. Amino acids, the building blocks of proteins, are required for adequate immune function.
- **Vitamins:** Specific vitamins are essential for immune function, including vitamin A (for maintaining epithelial barriers), vitamin D (for regulating immune responses), and vitamin E (an antioxidant that supports immune health). Vitamin C is also essential in some livestock species.
- **Minerals:** As mentioned earlier, zinc, selenium, and copper are required for various immune functions, including producing and adequately functioning white blood cells, acting as antioxidants, and supporting immune cell activity. Each mineral has a

particular effect on the metabolism and immune activity.
- **Probiotics:** Beneficial bacteria in the digestive tract, known as probiotics, can help improve gut health and indirectly strengthen immune function by maintaining a balanced gut microbiome. A healthy gut is essential for nutrient absorption and overall well-being.
- **Prebiotics:** These are dietary components that support the growth of beneficial gut bacteria. Prebiotics, such as certain types of fiber, can enhance the effectiveness of probiotics and promote a healthy gut environment. The pro and prebiotics combine to form synbiotics. These are combinations of probiotics and prebiotics. Synbiotics can provide a dual benefit by introducing beneficial bacteria with the nourishment they need to thrive.

Omega-3 Fatty Acids: These healthy fats have anti-inflammatory properties and can support the immune system by reducing inflammation.
- **Antioxidants:** Various antioxidants, like vitamin E and others, help protect immune cells from oxidative damage and support their proper functioning.
- **Water:** Adequate hydration is essential for all physiological processes, including the immune response. Water helps transport nutrients and remove waste products, supporting overall health and immunity.
- **Carbohydrates and Energy:** Livestock require energy from carbohydrates to fuel immune responses and maintain their health. Proper energy intake is essential for immune function.
- **Herbs and Botanicals:** Some plants, herbs, and botanicals have been used in livestock diets for their potential immunomodulatory effects. These include substances like echinacea, garlic, and oregano. These herbs should always be given in minute quantities and in consultation with a certified veterinarian.
- **Immunomodulators:** Certain compounds, like β-glucans derived from yeast or algae, have been shown to augment the immune system in livestock when included in their diets.

It's necessary to note that the specific dietary requirements for a robust immune system can vary among different livestock species and individual animals. To ensure optimal immune health, it's advisable to consult with a veterinarian or animal nutritionist to formulate a diet that meets the unique needs of the livestock in question.

Nutrition Deficiencies Affecting Immune System

Deficiencies in the vital nutrients and dietary components essential for a robust immune system in livestock can compromise immune function, leading to a range of health issues and increased susceptibility to diseases.

Protein Deficiency

Animals with insufficient dietary protein may experience reduced production of antibodies and immune cells. These antibodies are essential for recognizing and neutralizing pathogens, and immune cells play a crucial role in immune responses. A protein deficiency can weaken immune responses, making animals more susceptible to diseases. Their ability to combat infections may be compromised. Furthermore, slow wound healing and impaired tissue repair can also result from protein deficiency, as these processes need proteins to produce and repair damaged tissues.

Vitamin Deficiencies

Different vitamin deficiencies have varying effects. Vitamin A deficiency can impair the integrity of epithelial barriers, such as the skin and mucous membranes, which serve as the body's first line of defense against pathogens. Weakened barriers make it easier for infectious agents to enter the body. Likewise, vitamin D deficiency leads to improper regulation of immune responses. It can potentially result in autoimmune diseases, where the immune system attacks the body's tissues. Similarly, vitamin E deficiency weakens the immune system's ability to defend against oxidative stress and infections and increases disease susceptibility.

Mineral Deficiencies

White blood cells are vital for recognizing and attacking pathogens. A zinc deficiency reduces white blood cell function and impairs immune responses. Copper deficiency can also negatively impact the development and function of immune cells. This impairs the immune system's ability to defend against pathogens.

Probiotic and Prebiotic Deficiencies

An imbalance in the gut microbiome can result from deficiencies in probiotics and prebiotics. This condition, known as gut dysbiosis, can disrupt the immune system's interaction with the gut, increase the chances of developing gut infections, and make it challenging to protect against diseases.

Omega-3 Fatty Acid Deficiency

Omega-3 fatty acids have anti-inflammatory properties. A deficiency in these healthy fats can lead to improper immune regulation, potentially causing excessive inflammation and increasing the risk of chronic diseases. Livestock with omega-3 fatty acid deficiencies may exhibit reduced resistance to infectious diseases, as inflammation is a critical component of the immune response.

Antioxidant Deficiencies

A lack of antioxidants, such as selenium and vitamin E, can increase susceptibility to oxidative stress. Oxidative stress can damage immune cells and impair their function. Livestock with antioxidant deficiencies may weaken the immune system's ability to combat infections.

Carbohydrate and Energy Deficiency

Inadequate carbohydrate and energy intake can result in reduced energy levels, which can weaken the immune response. Immune cells require energy to function effectively. Animals with energy deficiencies may experience slower recovery from illness and compromised health as their bodies struggle to mount immune responses and repair damaged tissues.

Water Deficiency

Dehydration can obstruct the transport of nutrients and the removal of waste products, impacting immune cell function and the overall health of livestock. Likewise, an increased risk of heat stress can result from water deficiency, further compromising the immune system and leading to other health problems.

Herbs, Botanicals, and Immunomodulators

The effects of specific herbs, botanicals, and immunomodulators on immune function can vary. Some may strengthen the immune system, while others may have limited or unproven effects. The absence of these dietary components means missing the potential immunomodulatory benefits they could provide to livestock, potentially affecting their immune health.

Deficiencies in these vital nutrients and dietary components can significantly affect the immune system of livestock. These effects can manifest as increased susceptibility to diseases, compromised immune responses, slower illness recovery, and impaired overall health and productivity. Therefore, proper nutrition and dietary management are essential for maintaining a strong and effective immune system in livestock.

Gut Health and Nutrition

The gut, specifically the gastrointestinal (GI) tract, is a complex and dynamic system responsible for various critical functions, and nutrition significantly influences its health.

Providing Nutrients

Nutrition supplies the energy and essential nutrients required to grow, maintain, and repair the cells lining the GI tract. The cells in the gut mucosa have a rapid turnover, and proper nutrition ensures replenishment of these cells, keeping the integrity of the gut lining true.

Promoting a Healthy Gut Microbiome

The alimentary system is home to trillions of beneficial microorganisms, including bacteria, viruses, and fungi, collectively known as the gut microbiome. A balanced diet with appropriate fibers and prebiotics can support the growth of beneficial gut bacteria. These microorganisms are crucial in digestion, nutrient absorption, and immune function. A healthy gut microbiome helps protect against harmful pathogens and contributes to overall gut health.

Maintaining Gut Barrier Function

The gut lining is a barrier that prevents the entry of harmful substances, such as pathogens and toxins, into the bloodstream. Proper nutrition supports the upkeep of a strong gut barrier by providing the necessary nutrients for mucin production and tight junction proteins. A compromised gut barrier can lead to leaky gut syndrome, allowing unwanted substances to enter the bloodstream, potentially leading to inflammation and various health issues.

Modulating Inflammation

Certain nutrients and dietary components can cause or reduce gut inflammation. Omega-3 fatty acids, for example, have anti-inflammatory properties and help manage inflammatory conditions like inflammatory bowel disease (IBD). Proper nutrition can modulate the inflammatory response in the gut and contribute to gut health.

Preventing Gastrointestinal Disorders

Nutritional choices can impact the development and progression of various gastrointestinal disorders, including conditions like gastritis, gastroenteritis, and colorectal cancer. A fiber-rich diet with fruits, vegetables, and antioxidants can reduce the risk of some GI diseases and support gut health.

Balancing Gut pH

Nutrition can influence the pH levels in the gut. An optimal pH environment is vital for adequately-functioning digestive enzymes and the gut microbiome. Ph imbalances can lead to conditions like acid reflux, which dietary choices can influence.

Supporting Gut Motility

Adequate dietary fiber and hydration are needed to support healthy gut motility. Proper movement of food and waste through the GI tract prevents constipation and ensures the efficient absorption of nutrients.

Management of Food Allergies and Intolerances

Some animals may have food allergies or intolerances that affect their gut health. Proper nutrition, which includes avoiding trigger foods, can help manage these conditions and alleviate symptoms.

Livestock Disease Influences

Mastitis (Dairy Cattle)

Proper nutrition ensures that cows have the resources needed to maintain healthy udders, which can help prevent mastitis. Furthermore, a healthy cow is likelier to have a strong defense system that can respond effectively to intruding pathogens. When mastitis does occur, a robust immune response is necessary to fight off the infection and promote recovery.

Foot and Mouth Disease (Various Livestock)

In the case of foot and mouth disease, a well-nourished animal with a sound immune response is more likely to resist the infection and recover quickly. The virus can spread more easily among animals with weakened immune systems, emphasizing the importance of nutrition in preventing and managing this highly contagious disease.

Coccidiosis (Poultry, Cattle, Sheep, and Goats)
A balanced diet that meets the nutritional needs of livestock helps prevent coccidiosis. A well-nourished animal is better equipped to mount a strong immune response against coccidial parasites.

Respiratory Diseases (Swine, Poultry, Cattle)
Nutrients, including vitamins and minerals, support lung function and can help prevent respiratory diseases. Furthermore, a healthy immune system is critical for preventing and managing infections caused by respiratory pathogens like bacteria and viruses. Livestock with compromised immune systems are more vulnerable to severe respiratory infections.

Parasitic Infections (Various Livestock)
Parasites, such as gastrointestinal worms, can be particularly detrimental to animals with compromised immune responses. Adequate provision of probiotics through nutrition is essential for controlling and limiting the impact of parasitic infections.

Clostridial Diseases (Sheep and Cattle)
Proper nutrition supports their immune system, which is necessary for preventing and managing clostridial diseases. These diseases, caused by clostridium bacteria, can be particularly severe in animals with weakened immune defenses.

Mycoplasma Infections (Poultry, Swine, Cattle)
The nutrients fed to livestock strengthen the immune system, allowing it to develop a feasible immune response that decreases the severity and the duration of infections.

Salmonellosis (Various Livestock)
A healthy gut is less susceptible to colonization by salmonella bacteria, reducing the risk of salmonellosis. In addition, a robust immune system is vital for controlling the infection and preventing its spread to other animals. Proper nutrition and a strong immune response are key factors in preventing and managing Salmonella infections.

Management Practices to Follow

Access to Clean Water
Ensure a constant supply of clean, fresh water for your livestock. Adequate hydration is essential for overall health and immune function—regularly

clean water troughs and containers to prevent the growth of harmful bacteria and ensure good water quality.

Balanced Diet

Consult with a nutritionist or veterinarian to formulate a balanced diet for your livestock. Different species and life stages have unique nutritional requirements. Use high-quality, properly stored feed to ensure nutrient content and quality. Monitor feed availability, especially during extreme weather conditions, and adjust rations as needed.

Forage and Pasture Management

Practice rotational grazing to prevent overgrazing and allow pastures to recover. This approach can help maintain forage quality and reduce the risk of parasitic infections. Monitor forage quality and adjust the diet accordingly, especially during different seasons.

Vaccination and Disease Prevention

Put a vaccination program in place based on your livestock's specific disease risks. Consult with a veterinarian to develop a comprehensive vaccination schedule. You'll need proper biosecurity measures to prevent the introduction of diseases to your farm. Quarantine new animals to minimize the risk of disease transmission.

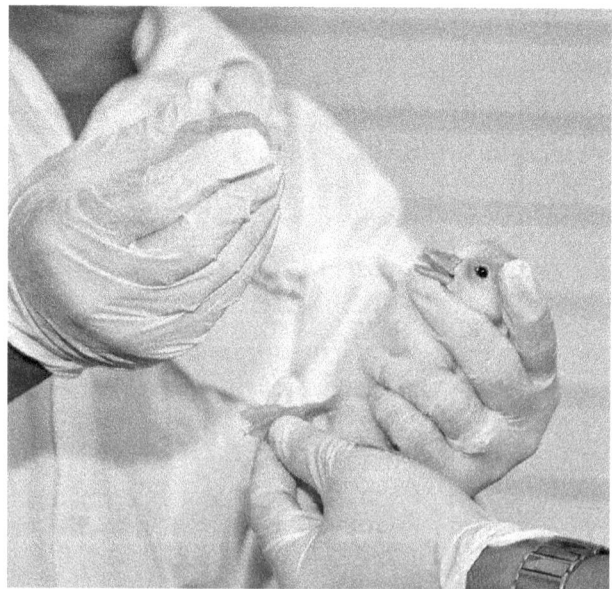

Put a vaccination program in place based on your livestock's specific disease risks.
https://commons.wikimedia.org/wiki/File:USAID_trains_animal_health_workers_in_poultry_vaccination_in_Vietnam._(5070816105).jpg

Stress Management

Minimize stress factors such as overcrowding, abrupt changes in diet, and transportation, as these can weaken the immune system. Handle livestock gently and calmly to reduce stress during routine management practices.

Environmental Hygiene

Keep living areas clean and well-ventilated to reduce the risk of respiratory infections. Properly manage manure and waste to minimize the risk of disease vectors breeding and keep your environment clean.

Mineral and Vitamin Supplements

Test your fields and forage for mineral content to determine if supplementation is necessary. Consult a nutritionist for specific recommendations. If necessary, set up mineral blocks or supplements.

Monitoring and Record Keeping

Regularly monitor the health and condition of your livestock. Look for signs of illness or stress. Keep detailed records of vaccinations, feeding practices, and health observations to identify trends and potential issues.

Parasite Control

Develop a parasite control program that includes regular deworming and rotational grazing practices. Use fecal egg counts to monitor parasite burdens and assess the effectiveness of your parasite control program.

Genetic Selection

When choosing breeding stock, consider genetic traits related to disease resistance and overall health. Select animals with a history of strong immunity and resistance to common diseases.

Temperature and Weather Considerations

Provide adequate shelter to protect livestock from extreme weather conditions. Exposure to cold, wet, or excessively hot environments can weaken the immune system. Adjust feeding schedules and quantities in response to seasonal changes in nutritional requirements.

Consultation and Education

Work closely with veterinarians, nutritionists, and extension services to stay informed about best practices and advancements in livestock management. Regularly educate yourself and your staff on the latest research and recommendations for nutrition and health management.

A balanced diet, including adequate protein, vitamins, minerals, and essential nutrients, is fundamental for immune function and overall well-being. Proper nutrition supports a robust immune system, reducing the risk and severity of diseases.

Components like probiotics, prebiotics, and synbiotics help maintain a healthy gut microbiome, while omega-3 fatty acids, antioxidants, and carbohydrates support immune responses. Deficiencies of these vital nutrients can compromise immune function and leave livestock vulnerable to infections and health issues.

Furthermore, practical management practices play a crucial role in maintaining optimal health. These include ensuring access to clean water, rotational grazing for forage quality, vaccination and disease prevention, stress reduction, environmental hygiene, and genetic selection for disease resistance.

Monitoring, record-keeping, and consultation with veterinarians and nutritionists are also vital. By incorporating these measures into livestock management, you can help ensure that your animals receive the right nutrition, maintain strong immune systems, and are better equipped to resist diseases, ultimately optimizing their overall health and productivity.

Chapter 6: Herbal Remedies

It is never easy seeing your livestock infected with parasites. Naturally, you want to find the best remedy for them, but traditional medicine contains chemicals that can be harmful to your animals and can do more harm than good with unwanted side effects. They are also more expensive, and it can take a while to see any real results. For this reason, you should consider natural options. Herbal remedies are a great alternative and much safer for your livestock.

This chapter explains herbal remedies and their historical use. It also presents some of the most common remedies with simple step-by-step instructions.

What Are Herbal Remedies?

Herbal remedies, also called botanical medicine, botanic therapy, and phytotherapy, are natural medicines made from different plant parts like stems, bark, flowers, roots, berries, and leaves. They are used to treat or prevent various diseases. Unlike regular medicines, herbal remedies aren't tested or regulated. Herbs are plants that are used for their savory, aromatic, and medicinal qualities.

Herbal remedies are among the most ancient treatments in the world.
https://pixabay.com/photos/natural-medicine-flower-essences-1738161/

Herbal remedies are among the most ancient treatments in the world. Most of the modern medicinal recipes are derived from folk medicine.

Historical Use of Herbal Medicine

People have used herbs and traditional livestock medicines for thousands of years. Herbs were extremely popular in many ancient cultures as they were used for different purposes. For instance, the Romans purified the air with dill.

One of the most renowned Greek physicians, Hippocrates, also found 400 herbs to treat diseases. Another Greek physician, Pedanius Dioscorides, wrote a book called "De Materia Medica" where he listed the benefits and medicinal uses of a variety of herbs. To this day, many people still use this book as a reference for natural medicine.

The ancient Egyptians also used plants as remedies. Archeologists found an ancient papyrus containing 700 medical formulas that were surprisingly advanced. They talked about using herbs like bayberry, basil, and aloe for medicinal purposes.

In the Middle Ages, people used herbs to preserve meat, cover rotten taste in food, and mask unpleasant body odors. Interestingly, herbal medicine wasn't popular at the time because it was associated with paganism and witchcraft. American Indians also knew about the significance of herbal remedies from their ancestors and grew different herbs in their gardens.

Herbalism reached its highest popularity from the 15th to the 17th centuries. Greek and Latin books were translated into English, and they were in high demand.

Throughout the years, herbalism faced many challenges with its association with witchcraft or others accusing it of being old wives' tales. However, things changed in the 20th century when the Council on Medical Education set standards for the medical use of herbs. However, many schools didn't put teaching herbs high on their list. Luckily, in the last 50 years, there has been a growing interest in folk medicine and herbalism, with many herbal remedies sold either online or in stores.

There isn't much evidence to show how people started using herbs as medicine or discovered which plants were safe for use and which were toxic. However, in 1 A.D., the Roman herbalist Pliny wrote that humans learned about safe herbs from animals like deer, dogs, and swallows.

Over the years, people have used herbs to cure their illnesses and treat their livestock and pets. Evidence from 60,000 years ago proves that humans used the same medicinal plants on themselves and their animals.

Herbs have been used to treat parasites in animals for centuries as well. Many ancient records showed people used extracts of garlic, castor oil, and areca nuts to control parasites in animals.

Growing Interest in Herbal Alternatives in Parasite Control

In the last few years, more and more people have been turning their attention to herbal remedies to treat their livestock. Natural medicine is better than traditional medicine in many respects. It doesn't cause many side effects, doesn't contain chemicals or artificial additives, is affordable, and doesn't cause irritation in the stomach.

Since you are raising livestock for meat and other products, you want to make sure these products are safe and clean. So, animals shouldn't contain any drug residue from medicines like anthelmintics or antibiotics, which can cause health problems to the people consuming their meat. Many farmers are treating their livestock with herbal remedies instead of chemical medicines to guarantee their livestock's safety and protect their family's health.

Are Herbal Remedies Effective?

Many people often wonder whether herbal remedies are effective on humans and animals. Well, if they weren't effective, would they still be popular worldwide after all these years?

In Asia and other countries, people use medicinal plants to treat parasites and worms in livestock. After using plants like Hedysarum coronarium and Lotus pedunculatus to treat their sheep, they noticed a 50% decrease in worm infestation. Tannin oil has also proved effective in weakening parasite activities in livestock.

According to a study published in Herald Scholarly Open Access, medicinal herbs with antibacterial activity, like turmeric, ginger, thyme, cinnamon, and cloves, can treat parasites in cattle.

In another study conducted by Cambridge University, many herbs with antiparasitic properties, like coconut oil, clove oil, anise, and goldthread, were shown to be effective in treating worms in farm animals.

Limitations of Herbal Remedies

Herbal remedies, like any type of medication, have their own limitations. Unfortunately, all the evidence that proves the effectiveness of herbal medicine is very limited. Although some scientific research backs them, their results and success are mainly based on traditional use.

A Cambridge University study also found that not all antiparasitic plants are effective. For instance, the neem tree effectively treats gastrointestinal diseases and other parasite-related issues in livestock. However, when its leaves were used to treat a sheep suffering from parasites, there was no anthelmintic effect.

A List of Herbal Remedies for Parasite Control

There are many types of medicinal plants in nature, and each has its own usage and benefits. This section explores herbal remedies with anthelmintic properties commonly used for parasite control. Anthelmintic plants kill intestinal parasites or eliminate them.

Aloe Vera

Although aloe vera is known for its soothing properties, it also has other benefits. The plant can eliminate or destroy parasites and protect the animals from further infections.

Active Ingredients:
- **Hormones**: They have anti-inflammatory properties and can heal wounds.
- **Fatty Acids**: They have anti-inflammatory properties.
- **Anthraquinones**: They have antiviral and antibacterial properties.
- **Sugars**: Provide the body with fructose and glucose.
- **Minerals** like zinc, sodium, potassium, manganese, magnesium, selenium, copper, chromium, and calcium.
- **Enzymes**: Reduce inflammation.
- **Vitamins** like choline, folic acid, and vitamins A, B12, C, and E.

Mechanisms of Action:
- **Antiseptic Effect**: Helps fight against viruses, bacteria, and fungi.
- **Antiviral Properties**: Boosts the immune system and protects it against infections.
- **Laxative Effects**: Increases mucus secretion and water connection in the intestines.
- **Anti-Inflammatory** properties.
- **Healing Properties**: Accelerates wound contraction.

Chamomile

Chamomile has anthelmintic properties and is high in tannin compounds. It paralyzes the parasite and reduces the egg-hatching process.

Active Compounds:
- Rutin
- Luteolin
- Apigenin
- Quercetin

These compounds have antioxidant and antibacterial properties.

Mechanisms of Action:
- Anti-cancer properties
- Anti-inflammatory properties
- Protects against cardiovascular issues
- Treats diarrhea

- Treats eczema
- Treats gastrointestinal conditions
- Boosts immune system

Coriander

Coriander seeds are well known for their healing properties. They are considered as fungicidal, bactericidal, and larvicidal. They can eliminate parasites and protect your body from these microorganisms.

Active Compounds:
- Linalool
- Terpinene
- Pinene
- Camphor
- Limonene
- Geranyl acetate
- Cymene
- Bornyl acetate
- Thymol
- Gallic acid

Mechanisms of Action:
- Anti-inflammatory properties
- Anti-cancer properties
- Relieves gastrointestinal discomfort
- Stimulates the liver to increase bile secretion

Garlic

Garlic is known for its antibacterial properties that can kill parasites in the gastrointestinal tract. It also has other health benefits, like improving the animals' immune systems and protecting them from new parasite infestations.

Garlic is known for its antibacterial properties that can kill parasites in the gastrointestinal tract.
https://pixabay.com/photos/garlic-ingredient-flavoring-3419544/

Active Compounds:

- **Organosulfur Compounds:** allicin, ajoene, alliin, allyl propyl disulfide, diallyl polysulfides, vinyl thins, diallyl tetrasulfide, diallyl sulfide, diallyl disulfide, diallyl trisulfide, and allyl methyl trisulfide.
- **Phenolic Compounds:** like sinapic acids, coumaric, hydroxybenzoic, vanillin ferulic, and caffeic.
- **Saponins:** like eruboside B and proto-eruboside B.

These compounds have antibacterial and antioxidant properties.

Mechanisms of Action:

- Antimicrobial properties
- Stimulates the immune system
- Protects against cardiovascular issues
- Anti-cancer properties
- Protects the digestive system

Ginger

Ginger is one of the most popular herbal remedies, and people have been using it for centuries to treat the digestive system. Although there isn't enough evidence to support its effect on parasites in livestock, there is no harm in experimenting with it. Add it to your animals' food every week and observe the results.

Active Compounds:
- Terpene compounds
- Phenolic, like paradols, shogaols, and gingerols
- Polyphenols like 10-gingerol and 8-gingerol

Mechanism of Action:
- Antimicrobial properties
- Antitumor properties
- Anti-inflammatory properties

Neem

People who used chemical remedies found neem a great and effective alternative. It doesn't only eliminate parasites but also protects livestock from fungal diseases. To protect their animals against parasites, farmers feed their livestock neem cake in some parts of India and use the plant's oils and leaves to prevent parasitic infections.

Active Compounds:
- Azadirachtin
- Quercetin
- Salannin
- Gedunin
- Sodium nimbinate
- Nimbidol
- Nimbidin
- Nimbin

Mechanisms of Action:
- Anti-inflammatory properties
- Anti-cancer properties
- Antioxidants properties

Pot Marigold

Pot marigold has antimicrobial properties and can be an effective treatment against any diseases caused by parasites or any other microorganisms.

Active Compounds:
- Lutein
- Sesquiterpene glycosides
- Saponins
- Triterpene glycosides
- Oleanane-type
- Triterpene oligoglycosides
- Flavonol glycoside

Mechanisms of Action:
- Antifungal properties
- Reduces inflammation
- Antioxidant properties

Pumpkin Seeds

Pumpkin seeds can be an effective treatment against tapeworms and other types of intestinal parasites. These seeds are high in amino acid cucurbits and can paralyze the parasites and expel them from the gastrointestinal tract.

Active Compounds:
- Cucurbitacins
- Tocopherols
- Phenolic compounds
- Unsaturated fatty acids
- Phytosterols
- Amino acids

Mechanisms of Action:
- Contains Omega-3 and omega-6 fatty acids
- Antioxidants properties
- Eliminates and protects from parasites

Tannin-Rich Plants

Tannin is a compound that can reduce or remove parasites in the digestive tract. Many herbs contain tannins, like sage, coriander, rosemary, mint, and licorice.

Active Compounds:
- Gallic acids
- D-glucose

Mechanisms of Action:
- Anti-inflammatory properties
- Healing wounds
- Antioxidant properties

Turmeric

Curcumin is one of the substances found in turmeric and has antiparasitic properties. It can also eliminate adult worms.

Active Compounds:
- Curcumin
- Volatile oil
- Curcuminoids

Mechanisms of Action:
- Anti-inflammatory properties
- Anti-cancer properties
- Antimicrobial properties

Wormwood

People have been using wormwood for centuries to kill human and animal parasites. They also strengthen the body's immune system and protect it against worms. This herb is extremely safe and has no side effects.

Active Compounds:
- Thujone
- Coumarins
- Flavonoids
- Phenolic acids
- Artemisinin
- Absinthin isomers

Mechanisms of Action:
- Aids in digestion
- Antifungal and antibacterial properties

- Protects the liver
- Anti-inflammatory properties
- Boosts the immune system
- Antioxidants properties
- Protect nerve cells
- Relieves pain

Making and Administering Herbal Remedies

Now that you understand the significance of herbal remedies, you are ready to discover various natural medicine formulations to treat your livestock.

Tea
Instructions:
1. Pick the leaves and roots of any of the plants mentioned here.
2. Leave them to dry in a warm and dark place.
3. After they dry, store them in labeled jars in a cool, dark place.
4. Take 1 teaspoon of dry leaves or roots and brew tea by adding hot water to them.

Strong Tea Brew
Instructions:
1. Simmer one teaspoon of plant parts in a stainless steel teapot.
2. Leave it to cool down, but don't remove the lid.
3. Use it while it is still fresh.

Put the tea in a syringe and administer it into the animals' mouths. You can give them this remedy once a day for a week. Make sure it's cooled down before they consume it.

Tincture #1
Instructions:
1. Cut or dice the plants' leaves or roots in a glass jar.
2. Pour in alcohol like vodka to cover the plant's parts, or mix the alcohol with water.
3. Cover the jar with a lid, label it, and add the date.
4. Store it for six weeks in a cool and dark place.

You can remove the plant material if you prefer and only store the liquid. If you don't want to use alcohol, use vinegar or glycerine instead. However, if you plan to store it for a long period of time, adding alcohol is the better option since it has a long shelf life.

Put the tincture in a dropper bottle and give your animal 10 to 15 drops once a day for a week.

Tincture #2
Ingredients:
- 1 1/4 cups of glycerin
- 2 smashed garlic cloves
- 1 tablespoon of oregano
- 1 tablespoon of thyme
- 1 tablespoon of Oregon grape root

2 tablespoons of echinacea leaves or roots **Instructions**:
1. Mix all the ingredients minus the glycerin.
2. Add the glycerin.
3. Put the mixture in a glass jar and seal it tightly.
4. Store the jar in a cool and dark place for a month.
5. Shake the jar once every day.
6. After one month, pour the tincture into a glass dropper bottle.

Administer by mouth by putting 10 to 15 drops in the animal's mouth or water.

Paste
Ingredients:
- 1 tablespoon of beeswax
- 1 tablespoon of dry herbs

2 tablespoons of olive oil **Instructions**:
1. Fill a jar with dry herbs, then pour the olive oil.
2. Put the jar in a dry and warm place and cover it with a cheesecloth to protect it from insects and dust.
3. Store it for six weeks to let the oil infuse.
4. Check the jar regularly for mold.
5. After the oil is infused, add the beeswax to create a paste.

Apply on your animal when needed.

Tonic

Ingredients:

- 2 cups of wormwood
- 2 cups of thyme leaf
- 2 cups of sage leaf
- 1 cup of rosemary leaf powder
- 1 cup of psyllium seed powder
- 1 cup of mustard seed powder
- 1 cup of ginger root powder
- 1 cup of garlic powder
- 1 cup of cayenne pepper powder
- 1 cup of black walnut hull powder
- 1 cup of anise seed powder (optional)
- 1/2 cup of powdered cloves
- 1 cup of cinnamon powder
- 2 cups of diatomaceous earth (Although it is a very effective remedy against parasites, it can make the tonic feel dusty, which can be hard on your animals' lungs. You can feed it to your animal separately to speed the healing process.)

Instructions:

1. Mix all the ingredients, then place them in a glass jar and label it.
2. Store in a cool and dark place.
3. Administer for a week, twice daily, and every six to eight weeks, or as needed.

N.B. Don't use black walnut or wormwood if your animal is pregnant. Equines should never consume black walnuts.

Herbal Dosage Ball

Ingredients:

- 1/2 cup of powdered herb
- 1/4 cup of flour (to hold the mixture together)
- 1/4 cup of honey or molasses

Instructions:
1. Mix the first two ingredients, then add the honey or molasses.
2. Knead them into a dough with your fingers or in a food processor.
3. Break it into 12 small balls, then coat each with the flour.

Only feed your animal one ball a day. You can feed it to them by hand, and they might eat it right away. If they don't, shove it into their mouth. In most cases, they will like it and swallow it. However, they might not and spit it out. If this happens, shove it further back. Break the ball into pieces for baby animals.

You can feed it to them by hand, and they might eat it right away.
https://commons.wikimedia.org/wiki/File:Feeding_the_sheep.jpg

Salves
Ingredients:
- 1 tablespoon of honey
- 1 tablespoon of beeswax
- 2 tablespoons of calendula-infused oil
- 2 tablespoons of oregano-infused oil
- 2 tablespoons of chamomile-infused oi.

Instructions:
1. Melt the beeswax and oils in a double boiler.
2. Then, pour them into a jar and leave them to cool down.
3. Once they cool down, add the honey to the jar.
4. Seal it tightly with a lid, and you can use it for 6 to 12 months.

Apply on your animal's infected area twice a day until they heal.

Bath Soap
Ingredients:
- 2 drops of eucalyptus essential oil
- 2 drops of tea tree essential oil
- 3 drops of oregano essential oil
- 3 drops of lavender essential oil

3 cups of Castille soap **Instructions:**
1. Mix all the ingredients and place them in a glass bottle.
2. Shower the animal with the soap once every two months.
3. Scrub and lather your animal well.

N.B. Some dosages may differ depending on the size and number of animals, so consult your vet to be safe.

Safety Concerns

Although herbal remedies are effective and beneficial, there are a few safety concerns you should be aware of to guarantee your livestock's safety.

Potential Toxicity

Not all herbs are safe, and some are extremely toxic for cattle. You should be aware of toxic herbs and avoid them to keep your livestock and family safe.

Water Hemlock

This is one of the most toxic and dangerous plants in North America. If your animal consumes a small amount, they can get very sick or even die.

Signs of Water Hemlock Poisoning:
- Excessive drooling
- Nervousness

- Fast heartbeat
- Muscle twitching
- Pupils dilation
- Rapid breathing
- Tremors
- Seizures
- Coma
- Death

Lupine

If a pregnant cow consumes lupine, her babies will most likely be crooked, possibly having skeletal defects or a cleft palate.

Signs of Lupine Poisoning:
- Depression
- Nervousness
- Convulsions
- Lack of muscle control
- Trouble breathing
- Lethargy
- Coma
- Death

Death Camas

Judging from the name, consuming this plant can be fatal for your livestock. So make sure to keep your animals away from it.

Signs of Death Camas Poisoning:
- Trouble breathing
- Nausea
- Vomiting
- Weak muscles
- Heart failure
- Lung congestion
- Coma
- Death

Poison Hemlock
Poison hemlock can also cause birth defects in piglets and calves.
Signs of Poison Hemlock Poisoning:
- Weak pulse
- Trembling
- Paralysis
- Depression
- Pupils dilation
- Convulsion
- Respiratory paralysis
- Bloody feces
- Coma
- Death

Nightshades
Nightshade can be toxic to poultry, sheep, swine, cattle, and horses.
Signs of Nightshades Poisoning:
- Trouble breathing with expiratory grunt
- Drowsiness
- Trembling
- Paralysis
- Progressive weakening
- Nasal discharge
- High temperature
- Skin turning yellow
- Distended gallbladder

Black Cherry
Signs of Black Cherry Poisoning:
- Trouble breathing
- Staggering
- Convulsion
- Anxiety
- Collapse

- Sudden death

Tips to Safely Use Herbal Medicine

It is better to be safe than sorry. Using natural medicine is tricky. One mistake can risk your animals' lives. Follow these tips so you can safely use herbal remedies.

- Learn about poisonous herbs and keep them away from your animals.
- When you are preparing a remedy at home, follow the instructions and the recommended dosage.
- If you buy herbal medicine from a store, check the ingredients, expiration date, side effects, and dosage before using.
- Only buy your herbs from a licensed herbalist.
- Learn everything about herbs, and don't hesitate to ask your vet if you have any concerns.
- Call the vet immediately if you give your animal herbal medicine, and they start showing any side effects.
- Watch out for allergic reactions. Call your vet if your animal has trouble breathing after you give him the medicine.
- If your animals take other medications, consult your vet before giving them herbal remedies.

Challenges with Using Herbal Remedies

Many people have encountered some challenges when using herbal remedies. For instance, some people used any type of herb, believing they were all the same. Naturally, they didn't see any real results. After some research, they realized there are specific herbs for treating parasites and started using them. Some livestock owners also didn't pay attention to the dosage guidelines and gave their animals high dosages, making them sick. After consulting their vets, they realized their mistake.

Some farmers found that their animals were getting sicker after the medicine. However, they later discovered that it was the medication's side effect. Some realized their mistake after losing their animals. Since then, they would always observe their animals after giving them medication to see if something was wrong.

You can't predict how your animal will react to an herbal remedy. They might get an allergy reaction, start showing side effects, or start getting better. Don't walk away from them after you give them the medication. Stay close for 20 to 30 minutes to observe their reaction. If everything seems fine and they aren't reacting strangely towards the remedy, then continue with the medication.

You and your family's safety depends on the health of your livestock. Consider treating them with natural remedies and stay away from harmful chemicals. Herbal medicine has existed since the beginning of time, and its popularity hasn't slowed down for a reason. It is effective and safe. However, if your animal isn't improving, you should consider traditional medicine. In some cases, it might be your only option.

Chapter 7: Additional Natural Strategies

Besides all the natural methods discussed in the previous chapters, there are additional natural approaches for holistic parasite management since integrating more than one measure is essential to achieve sustainable control of the parasites. This chapter explores these alternative strategies, based on scientific research and case studies, illustrating the successful application of those strategies in different livestock production systems.

Using Diatomaceous Earth

Due to a more than generous application of antiparasitic drugs in the past, parasites becoming resistant to artificial anthelmintics has become a growing concern for livestock owners. Diatomaceous earth (DE) can be a suitable alternative treatment for internal parasites because it doesn't lead to antiparasitic resistance. In goats and other ruminants, there is also the problem of weak innate and acquired immunity to ringworm infections, which can't be solved with other natural methods. Also known as Diatomite, diatomaceous earth, on the other hand, has been used for deworming livestock and other animals and even expelling GIT parasites from people's intestines for centuries. Besides being a dewormer, diatomaceous earth can also be used for insecticidal treatment, a filtering agent for air and water, supplements, and food additives. The minerals in this compound can also boost nutrition and growth in farm animals, leading to higher live weight gains and improved heat tolerance (particularly beneficial in sheep). Moreover, while it's primarily used for controlling internal parasites, it can also be effective against external ones like lice and fleas.

What Is Diatomaceous Earth?

Diatomaceous earth is a powder made from naturally occurring sedimentary rocks that are the remnants of fossilized algae.
SprocketRocket, CC0, via Wikimedia Commons:
https://commons.wikimedia.org/wiki/File:Diatomaceous_Earth.jpg

Diatomaceous earth is a powder made from naturally occurring sedimentary rocks that are the remnants of fossilized algae. A small quantity of diatomaceous earth powder contains millions of Diatoms, microscopic hard-shell algae that populated the earth millions of years ago. Deposited in seas and lakes, the shells turned into sediments, and when these water sources dried out, the result was a silicon-rich compound. Besides silicone, diatomaceous earth also contains other minerals in varying amounts, depending on its source.

How Does Diatomaceous Earth Work?

Several theories support how diatomaceous earth can help control and eliminate parasites from inside and outside the host's body. According to one, the effectiveness of this compound against intestinal parasites can be seen in the shape of the particles. They look like a cylinder riddled with holes and have a negative charge. As all those tiny cylinders move through the body, their negative charge attracts everything with a positive charge, which, besides heavy metals, also applies to the outer layer of many intestinal parasites and pathogens. The holes absorb everything the cylinders attract, trapping them inside the particles. Once trapped, the host has no problem eliminating them through the intestinal system. The good bacteria in the gastrointestinal tract have a neutral or negative charge, so they won't be affected by the particles – one of the major benefits of using diatomaceous earth. Moreover, eliminating the parasites from the digestive tract helps keep a healthy balance in the microbiome of good bacteria, improving the hosts' appetite and contributing to better weight gain. A healthy gut microbiome will also boost their immunity, rendering them less vulnerable to future infections and parasite infestations.

Another way diatomaceous earth acts against gastrointestinal parasites, specifically worms, is by breaking their life cycle. For example, several studies (Laing et al. 2013, Beltran and Martin 2015, Islam et al. 2016) showed that regular treatment with diatomaceous earth can prevent roundworm larvae from migrating from the dung to the herbage, where ruminants would consume it. Interrupting the parasite's life cycle this way ensures that the number of parasites on the pastures reduces over time, ultimately reducing the number of parasites carried by the host.

A similar theory correlates diatomaceous earth particles' small, sharp edges with their abrasive action on a coating of cysts and external parasitic insects. The particles scratch the waterproof surface of these animals' external coating and absorb lipids from their exoskeleton, eventually causing them to die. Whereas, yet another theory (Köster, 2010) claims that due to its unique composition, diatomaceous earth works together with digestive enzymes, acting as a buffer to create a hostile environment for gastrointestinal parasites, preventing them from feeding and reproducing.

According to a study searching for a less invasive alternative for worm control in pigeons (M. WIEWIÓRA et al., 2015), regularly giving the birds diatomaceous earth supplements significantly diminishes the number of parasites in the digestive tract. Scientists have confirmed the effectiveness of this approach on livestock, pets, and even people, which is why you can buy diatomaceous earth in different formulations (food-grade for humans, for various types of animals, etc.).

Benefits notwithstanding, some studies (Rahmann, G., & Seip, H., n.d.) show that while diatomaceous earth might affect parasite loads measured by fecal egg counts, it does not reduce some serious symptoms caused by internal and vector-borne parasites (for example, they found that anemia often remains even after the parasites have been expelled).

When using it for livestock parasite control, diatomaceous earth can be applied internally and externally. For example, if the animals are infested with fleas, powder specifically formulated for this purpose can be applied to their skin or fur. They must be dusted daily until the fleas die off and the animal's skin clears up. The treatment must be applied to all animals in the same group/location, regardless of whether they are affected. The fleas can migrate and hide within the fur of new hosts, so treating all animals is always recommended.

To apply diatomaceous earth against internal parasites, the powder should be dissolved in water and given to animals regularly (following the manufacturer's instructions). The process involves giving the supplement to the animals for several days until the amount of worms and eggs expelled through feces is reduced to the minimum acceptable limit. Use regular measurements and consult guides for acceptable minimums for different animals to combat parasitic infections and prevent their return.

Using Forage Plants and Trees with Bioactive Compounds

Plant material from foraged plants and trees contains natural bioactive compounds that deter or eliminate parasites from the host's body, making them an excellent resource for eliminating parasitic infections in animal husbandry. For example, the Nordic countries have a long and well-established history of using plants as natural anthelmintics for both animals and humans – and for a good reason.

How Can Plants Help Combat Parasitic Infestations?

Besides being nutrient-dense and helping meet animals' nutrient needs, plants are also packed with antiparasitic compounds called nutraceuticals. These are metabolites and secondary plant substances (compared to the compounds they are mainly used for and which provide nutritional value), like tannins, potent antiparasitic agents. Many plants contain tannins, but only those with high condensed tannins levels are suitable for controlling and combating parasitic infestations. These tannin-rich plants are known as bioactive forage but also have another benefit. Unlike plants used for medicinal purposes, which often have side effects and consequently have to be used carefully and with due precaution, tanniferous plants are non-toxic. If necessary, they can be applied in larger doses and across a long period. They can even be incorporated into the normal diet of livestock as a supplement to other plant materials used for feeding.

There are several theories on how condensed tannins help combat livestock parasites. According to a popular one, when they come in contact with the surface of the parasite's body, these compounds create a powerful reaction hindering the parasite's metabolic functions, food intake, reproduction, and mobility. According to another theory, tannin-rich bioforge also has an indirect way of acting against parasites. Supporters of this theory argue that when ingested, condensed tannins bind proteins to themselves, forming complexes that can withstand digestive degradation (particularly strong in ruminants). This way, proteins can get through the digestive tract, where they'll be absorbed. Some parasites, like nematodes, cause protein degradation and loss in the intestines, preventing the host from taking this essential nutrient. Because proteins are the building blocks of cells and essential for many metabolic processes, reduced protein intake results in impaired metabolic functions, one of which is immunity. By helping get more protein to the intestines, condensed tannins can balance protein levels and provide the host with much-needed resources for boosting their immunity and resilience to parasites and other intruders.

How to Use Tanniferous Plants

There are several options for integrating tanniferous plants into livestock diets. For example, they can be cultivated as arable crops and used in the normal feeding rotation as a preventive measure before an expected infestation. Or, they can be preserved as silage or hay and used, for example, for deworming later when infestation occurs. However, some studies suggest that using tanniferous plants this way might have negative consequences, too, including reduced feed intake or feed digestibility (Dawson et al., 1999), if the proper measures aren't followed. For example, not all tanniferous plants will provide enough nutrition or be suitable for digestion for all livestock types. According to another study (Coop and Kyriazakis, 2001), in small ruminants, these negative consequences are often outweighed by the positive impact of a parasite-free herd on productivity and economic gain.

Which Plants to Use for Parasite Control

Phytotherapy is a science-based prophylactic or therapeutic application of plants or bioactive compounds gained from plants for preventing or curing diseases. According to a study, this approach can be divided into traditional and allopathic phytotherapy (Hördegen, 2005). The former is based on generations of practices passed down through oral traditions (some of which are still used today and proven effective by modern science). The allopathic approach relies on scientific verification of anthelmintic plants or their bioactive components. This verification also considers possible risks and side effects, as is the case with the following description of plants with anthelmintic effects backed by science.

Trees and Shrubs

Trees and shrubs are particularly commonly used as a treatment against parasites. For example, Willow (Salix spp) has proven anthelmintic effects and anti-inflammatory actions, which are additional benefits when combating parasitic infections. Feeding willow leaves to livestock eliminates the worms and larvae from their body, while the bark can be effective against flukes. A decoction can be created from the bark and added to the animal's water to get the maximum benefits.

In some areas, their widespread availability is a vast reason shrubs and trees are the most viable solution for natural antiparasitic measures. For example, small ruminants like sheep often graze on a broad range of herbage (including trees and shrubs), so they'll likely take to beneficial ones, too. Increasing the variety of plants they can graze on is proven to be correlated with the sheep's improved resilience to parasitic nematodes (Diaz Lira et al., 2005).

Herbaceous Plants

A wide variety of herbaceous plants have been used against parasitic infections and have proven effective in reducing the harmful effects of parasite infestation in livestock. One of the most widely used natural anthelmintics is derived from the plant called goosefoot or American wormseed (chenopodium ambrosioides). Archeological records suggest that the oil derived from this plant has been used for several centuries. For example, in the 18th century, Swedish botanist Peter Kalm noted when visiting North America that European settlers in the American colonies and the indigenous inhabitants used chenopodium extract for treating ascaris (nematode) infections. It has since been proven that the principal ingredient in this plant is ascaridole, a terpene found in many other herbaceous plants, too – which are now cultivated as crops for grazing or to be preserved for later anthelmintic supplementation. Many herbs used as herbs have antiparasitic effects. Studies (Eminov, 1982.) found that many of these plants are effective against trichostrongylus larvae in sheep.

According to ancient Roman literature, plants of the Asteraceae family have also been used to treat parasitic infections. The Romans used dried, not-yet-opened flowers of Artemisia species to expel ascaris and tapeworms from animals and people. The active ingredient in these plants is santonin, which modern veterinary and pharmacological studies have widely researched. For example, researchers found that santonin has a neurotoxic effect on worms in low quantities, specifically targeting the ganglia of their nerve rings (Saunders Company, 1957). The common tansy (Tanacetum vulgare) is also a member of the Asteraceae family. Known for its active component, thujone, this plant is a widely used deworming agent in the Northern Hemisphere.

Other plants with widespread use against parasites are berries and common vegetables belonging to the onion family (allium spp.), the cabbage family (brassica spp.), and even carrots (Daucus carota). Likewise, in tropical regions, cucumber and pumpkin seeds are traditionally used for expelling tapeworms and larvae from the host's intestines. When the practice spread to other parts of the world, scientists proved that these plants contain cucurbitin, a potent antiparasitic agent. It's still considered one of the safest antiparasitic measures in non-ruminants across the world.

Besides its widespread use for smoking, the tobacco plant was also used for treating nematode infestations in tropical regions. Tobacco infusions were a common practice for keeping parasites away from ruminant livestock until the appearance of synthetic anthelmintics.

Pasture Plants

Probably the most effective way of incorporating plants to control parasitic infections is by using specialized crops for grazing. Due to this, this approach has attracted much attention and is currently the focus of researchers worldwide. So far, the most condensed tannins among pasture plants were found in legumes. For example, lotus major (lotus pedunculatus) is packed with proanthocyanidins, which is associated with reduced worm infestation in grazing lambs (Niezen JH et al., 1985). Quebracho has a similarly powerful anthelmintic effect and can kill adult nematodes and their larvae in several types of livestock.

Lichens and Ferns

One of the plants most commonly used in the traditional Nordic antiparasitic measure approaches is a fern called Dryopteris filix-mas. Moreover, the ancient Greeks also used this plant (specifically, a powder made from its rhizome) against tapeworms. Like this fern, other lichens and ferns also contain silicic acid, which acts as a potent anthelmintic agent.

Plant Compounds

Besides using whole plants or plant parts like leaves, seeds, etc., you can also use compounds containing active ingredients for treating parasitic infections. For example, garlic powder is an excellent dewormer and can be used as a livestock supplement. Similarly, you find ground-up mustard, fennel, carrot seeds, wild ginger, goosefoot, and pyrethrum in powder or oil form. Pyrethrum (made from chrysanthemum flower) is the most effective in powder form. Neem oil is a powerful insecticide extracted from the Indian neem tree and is another excellent alternative for ectoparasites.

To apply it externally and eliminate ectoparasites, regularly run the powder or oil along the animal's infested areas until the parasites are gone. Internal application should be based on the manufacturer's recommendations for each compound.

Another study (Rahmann, G., & Seip, H., n.d.) found the following plants to be effective in curing and controlling endo-parasite diseases in livestock:

- Chicory (cichorium intybus)
- Birdsfoot Trefoil (lotus corniculatus and lotus pedunculatus)
- Sulla (hedysarum coronarium)
- Sainfoin (onobrychis viciifolia)
- Quebracho (schinopsis spp.)
- Socarillo (dorycnium pentaphyllum)
- Chinese Lespedeza (lespedeza cuneata)
- Dock (rumex obtusifolius)
- Wattle leaves (acacia karroo)
- Heather (calluna vulgaris)
- Chestnut Tree (fruit – Castanea sativa)
- Common Dogwood (cornus sanguinea)

- Hazel tree (Corylus avellana)
- Erica (erica ssp)
- Pine tree (leaves – pinus sylvestris)
- Pomegranate (Punica granatum)
- Oak (Quercus spp)
- Black Locust (robinia pseudo acacia)
- Blackberry bush (rubus fruticosus)
- Willow (salix spp)
- Genista (leaves – sarothamnus scoparius)
- Grape Seed (extract – Vitis spp)

Selective Breeding Programs

Selective breeding programs rely on animals' inherent capacity to resist diseases they were previously not exposed to. As hosts, these animals can obstruct and alter the parasite's life cycle and become resistant to the illnesses the parasites carry or cause. Natural resistance is genetically coded, which means that by breeding resistant specimens, one can increase the resistance level in the next generation. Incorporating this genetic element into parasite and disease control in animal husbandry has many benefits. It creates a permanent change in the animal's genetic material, ensuring the consistency of the resistance and removing the need for further measures – which isn't the case with other natural parasite control methods that must be repeated or supplemented regularly. Moreover, through selective breeding, resistance to several diseases can be increased simultaneously.

Selective breeding can take many forms, depending on the available resources and the type of parasites. Some of the most popular techniques for achieving this goal include selecting specimens with the highest resistance levels for specific parasites and diseases, choosing the appropriate breeds based on the environment, and using crossbreeding to introduce genes into well-adapted species depending on the desired outcomes.

The method known as "marker-assisted selection," which involves identifying the biochemical, morphological, and genetic (DNA or RNA) markers connected to the degree of disease resistance, is frequently used to choose breeds and individual specimens. Since these markers are genetically associated with the characteristic, they serve as a trustworthy means of selecting the right animals for selective breeding. Screenable markers are linked to easily identifiable genotypes, while selectable markers are more suited for removing particular, typically undesired, genotypes. Since DNA may be reproduced in vitro and utilized as much as possible, marker-assisted selection offers a cost-effective method of selective breeding since it allows for assessing many markers with a single purified sample. This lowers the price of marking and selects breeds appropriate for the procedure.

Single nucleotide polymorphism detection is one technique that can be used to identify genetic markers. Molecular codes known as linked markers are found in close proximity to genes that encode resistance. For instance, ruminants have a number of phenotypic (physical) and genetic markers known to be encoded in genes located close to those that determine immunity to gastrointestinal parasites. Another option is to use the major histocompatibility complex (MHC), which contains many polymorphic genes that determine how animals react immunologically to infections and parasites. Two of the three MHC classes are linked to resistance in ruminants and belong to class 2.

Defining what traits to measure regarding resistance is another crucial question. One of the most popular ones is Fecal Egg Count (FEC), which is heritable – although its heritability varies depending on the animal breed and parasite species. Immune response evaluation is another useful trait to use for these purposes.

According to a study investigating the effects of genetic manipulation by selective breeding for improving resistance to gastrointestinal nematodes in sheep (Windon, R. G. 1990), this approach can be a reliable alternative for chemical parasitic control. This study relied on the genetic variation linked to nematode resistance in sheep, along with previous experiments proving the feasibility of creating breeds with a higher resistance level. The conductors of the study also considered immunity, seeing it as a major cause of host animals' resistance to parasites. Based on this, they also concluded that for selective breeding to act as a control measure against parasitic infestation, several things must happen:

- A reasonable heritability level must exist for the selected marker to maximize the response to selection.
- The method must be cost-effective when compared to other natural and artificial measures.

- Ideally, the found resistance should be non-specific – in other words, it acts against several different parasites.
- The selection must be based on a linked trait that doesn't require contact with the parasite to identify resistance.
- The process should not adversely affect livestock production in the next generation.
- The selection should ensure that if parasite adaptation occurs to the host's resistance, it will remain manageable with other measures.

While selective breeding can provide the advantage of using breed-inclusive variation for improving resistance to parasites, selection based on this criterion alone can have negative effects. For example, some breeds or individuals used for selective breeding might have high resistance but a low live weight gain. Along with the benefits of parasite control, the latter will also be emphasized in the next generation, leading to economic losses. Although selective breeding can eliminate other natural control procedures like pasture management, it can be used within an integrated approach to limit production losses and minimize costs.

Using Copper

Copper sulfate and copper oxide are naturally occurring substances often used as an antiparasitic measure. When ruminants ingest copper sulfate, it remains in their rumen, releasing trace particles of copper. These trace mineral particles interject into the parasite-host relationships, disrupting it and creating a hostile environment for the parasite. Studies (Bang et al., 1990) found that regular administration of copper compounds leads to a reasonable reduction of parasite number for some parasite species. Based on another study (Burke et al., 2004), the optimal dosage of copper compounds for expelling parasitic worms from lambs without causing toxicity is 0.07 ounces per dose. Mixing it with animal feed or drinking water is the easiest application of copper sulfate or copper oxide. It's recommended to follow the manufacturer's recommendation for the specific copper sulfate compound you're using.

Chapter 8: Livestock Parasites and Climate Change

Long-term changes in climatic conditions, precipitation, and temperature are brought about by climate change, which is also changing Earth's average weather patterns. This issue is mostly caused by greenhouse gases like carbon dioxide (CO_2), methane (CH_4), and nitrous oxide (N_2O) that are unnecessarily released into the atmosphere as a result of human activity. The term "global warming" refers to the steady rise in global temperatures brought on by these gases' ability to retain solar heat.

The global impacts of climate change extend far beyond just temperature fluctuations and are deeply intertwined with the dynamics of livestock parasites. Here's a closer look at the critical global effects of climate change and how they are connected to livestock parasites:

Rising Temperatures

This warming environment influences the distribution and survival of a range of livestock parasites.
https://pixabay.com/photos/climate-change-thermometer-3836835/

As global temperatures rise, so does the ambient temperature in many regions. This warming environment influences the distribution and survival of a range of livestock parasites. For example, some parasites that previously couldn't thrive in cooler areas may now find these regions more suitable, potentially exposing livestock to new parasitic threats.

Changing Precipitation Patterns

Traditional precipitation patterns change during climatic changes, leading to more frequent droughts and increased rainfall in some areas. These shifts in moisture availability affect the survival and propagation of parasites. For example, moisture-dependent parasites may thrive in regions experiencing more rainfall, while droughts diminish the availability of water sources needed for parasite development. During heavy rainfall and flooding, the dispersal of vector-borne parasites through mosquitoes and insects increases exponentially.

Biodiversity and Ecosystem Alterations

Changes in the climate disrupt the ecosystem, changing the distribution of wildlife and their parasites. As these changes occur, livestock may be exposed to new parasite vectors or reservoir hosts. Understanding these changes and gathering relevant information is crucial for effective parasite management.

Food and Water Resources

Changes in climate patterns bring changes to water and food availability for livestock. For example, in times of drought, water becomes scarce, and food supplies are limited, which can weaken the immune system, making livestock susceptible to parasitic infections. Likewise, extremely humid areas with persistent rainfall directly promote parasite dispersal.

Health Risks

Higher temperatures associated with climate change can influence the activity and distribution of disease vectors, which carry parasites and transmit diseases to humans and livestock. For instance, regions feasible for expansion of disease-carrying insects increase the risk of vector-borne parasites affecting livestock.

Economic and Livelihood Consequences

Climate change also leads to economic losses in the livestock industry due to reduced productivity, higher management costs for parasite control, and increased veterinary expenses. These economic consequences have a cascading impact on the livelihoods of individuals and communities dependent on livestock for income and sustenance.

Migration and Conflict

As climate change renders certain regions less habitable, it leads to population displacement, including livestock. Displaced livestock could carry parasites to new areas, potentially introducing novel parasitic challenges.

Although these challenges can become a nuisance when not dealt with properly, most factors that potentially increase the incidence of parasitic infections in livestock can be avoided by implementing strict quarantine and monitoring protocols.

The global impacts of climate change are intricately linked with the dynamics of livestock parasites. Temperature, precipitation, and extreme weather events influence the prevalence and distribution of these parasites. As explained earlier, addressing the health and productivity of livestock in a changing climate requires a comprehensive understanding of these interconnections and implementing adaptive and mitigative strategies for effective parasite management.

Parasite Life-Cycle Influences

As you already know, temperature, humidity, and precipitation affect the survival, development, and transmission of parasites, ultimately impacting the health and productivity of livestock. Here's how these factors can influence the life cycles of different parasites.

Temperature

Warmer temperatures accelerate the development of many parasites. For example, gastrointestinal nematodes like haemonchus contortus in sheep and cattle thrive in warm and humid conditions. These nematodes can quickly complete their life cycles in warm climates, leading to more frequent infections.

On the flip side, cold temperatures can slow down or halt the development of some parasites. Liver flukes (Fasciola hepatica), which infect the liver of cattle and sheep, are less active in cold conditions. As a result, their transmission is reduced in colder climates.

Humidity

High humidity creates favorable conditions for the survival of many external parasites. The poultry red mite (dermanyssus gallinae) is a blood-feeding ectoparasite that infects chickens. As high humidity is essential for the mite's survival, it leads to infestations in poultry houses, causing stress and reduced egg production.

On the other hand, low humidity can desiccate and kill certain parasites. For example, the eggs and larvae of some gastrointestinal nematodes are susceptible to desiccation. In arid regions with low humidity, limited transmission of these parasites is reported, further confirming the effects of humidity on parasites.

Precipitation

Increased rainfall creates breeding sites for parasites and their vectors. The stable fly (stomoxys calcitrans), which feeds on the blood of cattle, loves wet, decaying organic matter to breed in. Conversely, drought can reduce the availability of water sources, affecting livestock drinking behavior and potentially increasing the risk of waterborne parasites. For example, liver flukes require a freshwater snail as an intermediate host. In drought conditions, these freshwater snails fail to survive, disrupting the fluke's life cycle and limiting the chances of developing waterborne parasitic infections.

Specific parasite control measures you can take include adjusting deworming schedules, implementing vector control measures, and providing appropriate shelter and management practices to mitigate the impact of environmental factors on parasite transmission and livestock

health.
Temperature-Dependent Parasites

Although mentioned earlier, it's an evident change that influences parasite distribution. Warmer temperatures associated with climate change enable the movement of temperature-sensitive parasites to higher altitudes and latitudes. For example, parasites previously limited to lower elevations may now find the climate suitable for survival and reproduction at higher altitudes.

- The liver fluke Fasciola hepatica, which primarily affected lowland areas, has been reported in higher altitudes and latitudes as temperatures have increased.
- The spread of ticks carrying Lyme disease (Ixodes scapularis) to more northern regions in North America as temperatures have risen.

Altered Ecosystems and Host Distribution

Human activity, habitat loss, and land use change can alter ecosystems, impacting parasite distribution. The loss of natural habitats disrupts ecological balances, leading to modification in host populations and consequently affecting parasites.

- Deforestation increases contact between humans and wildlife, potentially transmitting zoonotic parasites from wildlife to humans.
- Changes in agricultural practices may result in shifts in livestock populations, which can influence the distribution of livestock parasites.

Invasive Species and Trade

The global movement of animals and goods can introduce new host species and their associated parasites to new regions.

- Introducing invasive species like the Asian tiger mosquito (Aedes albopictus) to new regions has increased the risk of diseases like dengue and chikungunya in previously unaffected areas.
- The international livestock trade can introduce parasites not previously found in a region, impacting the local parasite landscape.

Human Activities and Infrastructure

Urban areas create microclimates and environmental conditions conducive to certain parasites and their vectors. Pollution and the presence of human-made water sources can provide breeding sites for disease vectors, causing adjustments to parasite distribution, such as:

- The proliferation of culex mosquitoes in urban environments contributed to the spread of the West Nile virus.
- The increased distribution of snail-borne diseases in areas with polluted water sources.

Veterinarians, public health officials, and ecologists need to monitor and adapt to these changes to prevent the spread of diseases and mitigate their impact on human and animal health. This may involve modifying vaccination protocols, implementing vector control measures, and enhancing surveillance and diagnostic capabilities to manage changing parasite landscapes effectively.

Emerging Parasitic Infections

Emerging parasitic diseases have either newly appeared or have significantly increased in incidence, geographic range, or host range. These diseases pose challenges for identification and management due to various factors.

Leishmaniasis

Leishmaniasis is caused by protozoan parasites of the leishmania genus and is transmitted through sandfly bites. This disease has shown signs of emergence in new regions, possibly due to climate change and human migration.

Leishmaniasis can manifest in various clinical forms, including cutaneous, visceral, and mucocutaneous. The diversity in its clinical presentation can make diagnosis challenging. Its symptoms may overlap with other diseases, leading to misdiagnosis and delayed treatment. Access to accurate diagnostic tools is limited in resource-limited regions, hindering timely diagnosis and treatment.

Chagas Disease

Chagas disease is caused by the parasite Trypanosoma cruzi and is primarily transmitted by triatomine bugs. It has extended beyond its traditional boundaries in Latin America. Chagas disease often remains asymptomatic for years, making early diagnosis difficult. Diagnostic tests for Chagas disease may have limited sensitivity and specificity, leading to false negatives. The triatomine bug is complex and requires sustained efforts to control the disease's vector.

Toxoplasmosis

Toxoplasmosis, caused by the parasite Toxoplasma gondii, has been recognized as an emerging disease in some regions. Many infections remain asymptomatic, making it challenging to identify and manage. T. gondii can infect many animals, making control and prevention complex. Routine screening for toxoplasmosis in pregnancy is not universally implemented, potentially missing cases in pregnant women.

Angiostrongyliasis

Angiostrongyliasis, caused by the parasitic nematode Angiostrongylus cantonensis, has also emerged in new regions. Lack of awareness: Many healthcare providers need to become more familiar with the disease. A lack of experience with it results in misdiagnosis or delayed diagnosis. Clinical manifestations can range from mild headache and nausea to severe neurological complications, making diagnosis challenging. There are no specific antiparasitic drugs for angiostrongyliasis, and management is primarily supportive.

Babesiosis

Babesiosis is caused by intraerythrocytic parasites of the babesia genus and is transmitted through tick bites. Cases have been emerging in new regions. Babesiosis mimics malaria symptoms, leading to misdiagnosis. The disease can range from mild flu-like symptoms to severe, life-threatening conditions, making it difficult to predict disease outcomes. Some babesia species have zoonotic potential, complicating the understanding of disease dynamics.

Dirofilariasis

Dirofilariasis is caused by filarial nematodes, particularly Dirofilaria immitis, and is transmitted by mosquitoes. The disease has expanded into new regions. Many infected individuals remain asymptomatic; the disease may only be discovered incidentally.

Altered climate conditions have expanded the habitat range of the disease vectors, increasing the risk of transmission to humans and pets. Dirofilariasis can be misdiagnosed as other respiratory conditions due to its diverse clinical manifestations.

Challenges in Identifying and Managing Emerging Parasitic Diseases

- **Diagnostic Limitations**: Many emerging parasitic diseases present with diverse or nonspecific symptoms, and diagnostic tools may lack sensitivity and specificity.
- **Globalization and Travel**: Increased travel and global trade facilitate the movement of parasites and disease vectors across borders.
- **Vector-Borne Diseases**: Controlling parasites transmitted by vectors (e.g., mosquitoes, ticks) can be challenging, especially when the vectors adapt to new environments.
- **Environmental Changes**: Alterations in ecosystems, including climate change, can impact parasite distribution and transmission dynamics.
- **Drug Resistance**: The emergence of drug-resistant strains can limit treatment options and increase the difficulty of disease management.
- **Zoonotic Transmission**: Many emerging parasitic diseases are zoonotic, creating complex epidemiological patterns and disease control challenges.
- **Limited Awareness**: Healthcare providers, public health systems, and communities may lack awareness of these emerging diseases, leading to delayed recognition and response.

Addressing these challenges requires a multifaceted approach, including improved surveillance, research, diagnostic tools, and public health education. It is crucial to remain vigilant and adapt to these emerging parasitic diseases to prevent their further spread and mitigate their impact on human and animal health.

Eco-Friendly Strategies

Adapting to climate-driven changes in parasite dynamics is paramount for sustaining livestock health and ensuring the resilience of agricultural systems. As climate change alters temperature and precipitation patterns, parasites that affect livestock are shifting in distribution, intensity, and seasonality. A range of strategies can be employed to address these challenges, each designed to mitigate the risks associated with evolving parasite dynamics. These strategies encompass monitoring and surveillance, breeding and genetics, pasture and grazing management, deworming practices, biosecurity, nutrition, parasite-resistant forages, and more. This comprehensive overview explores each of these strategies in detail.

Improved Monitoring and Surveillance

- **Regular Assessment:** Consistent monitoring and surveillance programs are pivotal. Regularly collect samples and data to assess the prevalence and intensity of parasitic infections in livestock populations.
- **Advanced Diagnostic Tools:** Use advanced diagnostic techniques like PCR-based tests and serological assays. These tools offer increased sensitivity and specificity in detecting parasitic infections, enabling more accurate and early diagnosis.

Climate-Resilient Breeding

- **Selective Breeding:** Select livestock breeds that exhibit resilience to climate-related stressors and parasitic infections. Breeding for resilience improves the overall health and productivity of livestock.
- **Resistance Traits:** Include resistance traits in your breeding objectives. These traits can reduce the reliance on chemical treatments and the associated risks of drug resistance.

Grazing Management

- **Rotational Grazing**: Implement rotational grazing practices to optimize pasture health and minimize parasite contamination. Regularly moving livestock to new pastures can break the parasite life cycle.
- **Adapted Schedules**: Adjust grazing schedules and stocking rates based on climate-driven changes in forage availability and parasite risk. This adaptive approach reduces the risk of overgrazing and pasture degradation.

Strategic Deworming

- **Informed Decision-Making:** Develop deworming strategies based on the data you've collected from monitoring your programs and environmental conditions. Avoid routine deworming and instead adopt targeted approaches.
- **Reduced Overuse:** Reducing unnecessary deworming helps mitigate the risk of developing drug resistance in parasite

populations. Deworm only when indicated by monitoring results.

Biosecurity Measures
- **Preventing Introduction:** Establish strict biosecurity measures to prevent the introduction of new parasite strains or species through animal movements. Effective biosecurity includes quarantine protocols for incoming animals.
- **Isolation and Quarantine:** Isolate and quarantine newly introduced animals to prevent the spread of parasites. This practice minimizes the risk of disease transmission within the herd.

Nutritional Management
- **Dietary Adaptations:** Adjust livestock diets to address climate-induced nutritional deficiencies and support the immune system. Balanced nutrition helps animals cope with the stressors associated with parasitic infections.
- **Supplementary Nutrition:** To maintain livestock health and resilience, provide supplementary nutrition during periods of reduced forage availability, such as during droughts.
- **Forage Selection:** Select and plant forage varieties that resist certain parasites naturally. These forages may contain compounds that reduce the incidence of parasitic infections in livestock.

Alternative Treatments and Preventatives
- **Exploring Alternatives:** Explore alternative treatments and preventatives such as herbal remedies or biological control agents. These options can complement or replace traditional chemical treatments.
- **Parasitic Vaccines:** Consider using parasitic vaccines where they are available and appropriate for the specific parasites affecting livestock.
- **Anti-Parasitic Plants:** Certain plant species contain secondary compounds with anti-parasitic properties. For example, some tannin-rich forage plants can inhibit parasite development in the gut.
- **Parasite Control:** Encouraging livestock to consume these anti-parasitic plants can reduce parasite exposure and lessen the need for chemical treatments.

Integrated Pest Management (IPM)
- **Holistic Approach:** Implement an integrated pest management (IPM) approach that combines various strategies, including biological control methods (e.g., nematophagous fungi), targeted chemical treatments, and rotational grazing.
- **Environmental Considerations:** Prioritize IPM practices that minimize environmental impacts and preserve beneficial organisms while controlling parasite populations.

Research and Innovation
- **Scientific Investigation:** Invest in research to understand the specific impacts of climate change on local parasite dynamics. Research can inform adaptation strategies and improve their effectiveness.
- **Innovative Solutions:** Promote innovative solutions like developing climate-resistant livestock breeds and novel treatment methods that account for changing environmental conditions.

Collaboration and Knowledge Sharing
- **Stakeholder Collaboration:** Collaborate with local agricultural extension services, universities, and research institutions to access the latest information on parasite management in a changing climate.
- **Peer Exchange:** Share knowledge and experiences with other livestock producers and researchers to benefit from collective wisdom and best practices in parasite management.

Climate-Resilient Infrastructure
- **Livestock Housing:** Adapt livestock housing and infrastructure to cope with extreme weather events and temperature fluctuations. Ensure that animals have access to shade and proper ventilation during heatwaves and other climate-related stressors.

Diversification of Livestock Species
- **Species Variation:** Consider diversifying livestock species to adapt to changing conditions. Some species may be more resilient to certain parasites or better suited to altered environmental conditions, reducing the risk associated with monospecific farming.

- **Awareness:** Develop awareness of sustainable and climate-resilient practices while advocating for their adoption among stakeholders and the wider agricultural community.

Heat Stress Mitigation

Rising temperatures due to climate change can subject livestock to heat stress, which weakens their immune system, making them more susceptible to parasitic infections. To mitigate heat stress, ensure livestock can access shade, such as natural tree cover or purpose-built shelters. Adequate ventilation is crucial to prevent overheating. Nonetheless, providing access to cool, clean water sources helps maintain hydration and regulate body temperature.

Genetic Resistance

- **Breeding for Resistance:** Genetic resistance is selectively breeding livestock for inherent immunity or resistance to specific parasites.
- **Selective Breeding Programs:** Breeding programs can be established to develop livestock with natural resistance traits. For example, selecting animals with higher red blood cell counts may reduce susceptibility to blood-feeding parasites like haemonchus contortus.
- **Sustainable Parasite Control:** Genetic resistance can complement other parasite control strategies and reduce the reliance on chemical treatments.

Soil Improvement

- **Soil-Forage-Parasite Connection:** The health of the soil directly influences forage quality and quantity, which, in turn, affects livestock grazing patterns and susceptibility to parasitic infections.
- **Reseeding and Soil Amendments:** Practices like reseeding pastures with high-nutrient forages and soil amendments can improve soil health and forage quality. Healthy soils promote the growth of nutritious forage that supports livestock health and resilience.
- **Reduction in Parasite Exposure:** High-quality forage helps reduce the risk of parasitic infections by providing essential nutrients and supporting the well-being of livestock.

Remote Sensing and GIS

- **Environmental Data:** Remote sensing technologies, including satellite imagery and drones, can provide detailed environmental

data.
- **Geospatial Analysis:** Geographic Information Systems (GIS) are used to integrate climate data, soil quality, and vegetation cover. Such analysis can predict disease risk and help you to plan grazing strategies based on climate data.

Integrating climate and environmental data into livestock management decisions allows for informed, data-driven approaches that reduce the risk of parasitic infections.

Seasonal Management

- **Adaptation to Climate Shifts:** Climate change causes seasonal shifts, impacting forage availability and parasite dynamics.
- **Breeding and Weaning:** Adapt livestock breeding and weaning schedules to align with climate-driven changes. This can help optimize the health and well-being of animals based on the timing of forage availability and parasite risk.

Flexibility in management practices is crucial to adapt to changing seasonal patterns and the associated challenges.

Climate-Resilient Livestock Shelters

- **Extreme Weather Protection:** Climate-resilient livestock shelters are designed to protect animals from extreme weather events, such as hurricanes, heavy rainfall, and extreme temperatures.
- **Stress Reduction:** Adequate shelter can reduce the stress experienced by livestock during adverse weather conditions. Stress reduction supports the immune system and decreases susceptibility to parasitic infections.

Shelter design should consider local climate conditions and the specific needs of the livestock species to be housed.

These comprehensive adaptation strategies address the multifaceted challenges climate-driven changes in parasite dynamics pose. Combined, they can contribute to the long-term health and sustainability of livestock systems in a changing climate. Collaboration among stakeholders and ongoing research is vital for tailoring these strategies to specific regions and ensuring their effectiveness.

Glossary of Terms and Parasite Reference

A – Z Glossary

- **Acquired Immunity –** Acquired immunity refers to the immune system's ability to recognize and remember specific pathogens, like parasites, after an initial exposure. This recognition allows the immune system to respond more effectively at subsequent encounters with the same pathogen.
- **Adaptation –** In the context of natural parasitic control, adaptation refers to the evolutionary process through which parasites and their hosts develop specific traits and behaviors that help them survive and reproduce in their respective environments. Parasites may adapt to exploit host resources or evade the host's immune system, while hosts may develop adaptive responses to resist parasitic infection. This ongoing adaptation process often leads to coevolution between parasites and their hosts.
- **Antagonistic Coevolution –** Antagonistic coevolution is a dynamic evolutionary relationship between parasites and their hosts in which each exerts selective pressure on the other. Parasites evolve strategies to exploit hosts, while hosts evolve defenses to resist or tolerate parasite infection. This constant back-and-forth competition drives the development of new traits and

countermeasures in both parasites and hosts, resulting in an evolutionary arms race.

- **Antiparasitic Compounds** – Antiparasitic compounds are substances, often chemical or biological, used to target and eliminate parasitic infections. These compounds can include medications, such as antiparasitic drugs or herbal remedies, designed to kill or inhibit the growth of parasites. They are crucial for the treatment and control of parasitic diseases in both humans and animals.
- **Antiparasitic Drugs** – Antiparasitic drugs are pharmaceutical substances specifically formulated to treat parasitic infections in humans and animals. These drugs may target various types of parasites, including protozoa, helminths (worms), and ectoparasites like ticks and fleas. They work by disrupting the parasites' life cycles, metabolism, or reproductive processes, ultimately leading to their elimination from the host's body.
- **Behavioral Fever** – Behavioral fever is a phenomenon observed in some host species as a response to parasitic infection. When infected by certain parasites, hosts may exhibit an increase in body temperature as a deliberate behavioral response to combat the infection. This elevated temperature can create an environment less conducive to the parasite's survival and reproduction, contributing to the host's defense mechanisms.
- **Biological Control** – Biological control is a method of pest management that involves using natural enemies, like parasitoid wasps, predators, or pathogens, to regulate populations of pest species. In the context of natural parasitic control, biological control refers to using one parasite or organism to control another, often to reduce the impact of harmful parasites on ecosystems or agriculture.
- **Coevolution** – Coevolution is the simultaneous and reciprocal evolution of two or more species that interact closely with each other, like parasites and their hosts. In this process, each species exerts selective pressure on the other, developing characteristics and adaptations that allow them to exploit or defend against one another.
- **Commensalism** – Commensalism is a kind of symbiotic interaction in which the relationship between two species benefits

one of the species while causing little to no harm or advantage to the other. When discussing interactions that result in one organism residing in a host without harming or benefiting from it in a way that lowers the population of harmful parasites, the term "commensalism" may be used concerning natural parasitic control.
- **Ectoparasite** – An ectoparasite is a type of parasite that lives on the external surface of its host. These parasites feed on the host's blood, skin, or other bodily fluids. Examples of ectoparasites include ticks, fleas, lice, and certain mites.
- **Eco-Epidemiology** – Eco-epidemiology is a field of study that examines the ecological and epidemiological factors influencing the transmission and spread of infectious diseases, including those caused by parasites.
- **Endoparasite** – An endoparasite is a parasite that lives inside the body of its host. These parasites can inhabit various internal organs or tissues and feed on the host's blood, tissues, or bodily fluids. Examples of endoparasites include tapeworms, liver flukes, and some protozoan parasites.
- **Entomopathogenic Fungi** – Entomopathogenic fungi are a group of fungi that are pathogenic to insects. They infect and kill a number of insect species and are often used as biological control agents to manage insect pests.
- **Host** – In the context of parasitism, the host is the organism that harbors and provides resources for the parasite. Hosts can be plants, animals, or even microorganisms, and they may experience harm or negative effects due to the parasite's presence.
- **Hyperparasitism** – Hyperparasitism is a form of parasitism in which one parasite is parasitized by another. In this relationship, the primary parasite, which is already living in or on a host, becomes the host for a secondary parasite. This secondary parasite may target the primary parasite for resources or reproduction.
- **Immune Response** – The immune response is the collective set of physiological and biochemical mechanisms that organisms, including hosts, use to defend against foreign invaders, such as parasites. When a parasite infects a host, the host's immune

system initiates a response to recognize, neutralize, and eliminate the parasite. The nature of the immune response varies depending on the type of parasite and the host species.

- **Immunoparasitology** – Immunoparasitology is a branch of parasitology that focuses on the study of host immune responses to parasitic infections. It investigates how hosts recognize and combat parasitic invaders, the development of immune memory, and the mechanisms underlying resistance or susceptibility to parasitic diseases.
- **Integrated Pest Management** – Integrated pest management (IPM) is an ecological pest and parasitic control approach. It involves using multiple strategies, including biological control, cultural practices, and chemical treatments, to manage pests and parasites in agriculture, horticulture, and forestry while minimizing the environmental impact.
- **Intermediate Host** – An intermediate host is a host species that a parasite uses to complete a stage in its life cycle but not the sexual phase. Parasites often require multiple hosts in their life cycle, with the intermediate host serving as a transitional stage for the parasite's development before reaching the definitive host for reproduction.
- **Microbiome** – The microbiome is the community of microorganisms, including bacteria, fungi, viruses, and other microbes, that inhabit the body of a host organism. These microorganisms can play a crucial role in host-parasite interactions by influencing the host's immune response and overall health. Understanding the microbiome is essential for studying natural parasitic control.
- **Mutualism** – Mutualism is a type of symbiotic relationship between two species in which both partners benefit from the association. In the context of parasitic control, mutualistic interactions can occur when one organism, such as a microbe, helps the host resist or tolerate parasitic infections, leading to a mutually beneficial outcome for both parties.
- **Parasite Manipulation** – Parasite manipulation is a phenomenon where parasites, through various mechanisms, alter the behavior or physiology of their host organisms to promote their own survival and reproduction. This can include changes in host

behavior that benefit the parasite, such as increased predation risk, ultimately aiding the parasite in completing its life cycle.
- **Parasitism** – Parasitism is a type of symbiotic relationship in which one organism (the parasite) benefits at the expense of another organism (the host). Parasites derive resources from the host, potentially causing harm or reducing the host's fitness. Parasitism is a fundamental concept in the study of natural parasitic control.
- **Phoresy** – Phoresy is a type of commensalism in which one organism uses another organism as a means of transport to a new location. In the context of natural parasitic control, phoresy may involve parasites or organisms that attach themselves to a host and use it to disperse to new environments.
- **Predation** – This is a biological interaction in which one organism (the predator) captures, kills, and consumes another organism (the prey) for food. Predation can play a role in controlling the populations of organisms, including parasites, within ecosystems.
- **Predatory Nematodes** – These are microscopic roundworms that feed on other organisms, including harmful parasitic nematodes. They are used in biological control to manage parasitic nematodes that affect plants, animals, and humans. Predatory nematodes help reduce the population of harmful parasites by preying on them.
- **Reservoir Host** – A reservoir host is a host species that carries and maintains a parasite, often serving as a source of infection for other hosts, including humans. Reservoir hosts can be important in the epidemiology of parasitic diseases, as they play a role in the persistence and transmission of the parasite.
- **Symbiosis** – Symbiosis is a broad ecological term that refers to the interaction between two different species that live together in close proximity. In the context of parasitic control, symbiotic relationships can include mutualism, commensalism, and parasitism, depending on the effects and benefits of the species involved.
- **Transmission** – Transmission refers to the process of passing parasites from one host to another. This can occur through various means, such as direct contact, ingestion, or vector-

mediated transmission. Understanding the mechanisms of transmission is essential for controlling parasitic infections.
- **Trophic Cascade** – A trophic cascade is an ecological phenomenon where changes in the abundance or behavior of one species in an ecosystem result in cascading effects on other species at different trophic levels. In the context of natural parasitic control, trophic cascades may affect the population dynamics of parasites, their hosts, and their interactions.
- **Vector** – A vector is an organism, often an arthropod such as a mosquito or tick, that can transmit parasites or pathogens from one host to another. Vectors play a critical role in transmitting various parasitic diseases, including malaria, Lyme disease, and others.
- **Zoonosis** – Zoonosis is a disease that can be transmitted from animals to humans. Many parasitic diseases have zoonotic potential, and understanding the dynamics of zoonotic parasites is crucial for both animal and human health.

Common Livestock Parasites (and How to Deal with Them)

- **Gastrointestinal Nematodes (Roundworms):** Implement regular deworming schedules using appropriate anthelmintic drugs and practice pasture management to reduce exposure to infective larvae.
- **Coccidia:** Improve sanitation in livestock housing, provide clean water sources, and consider using coccidiostats in feed or water to control coccidiosis.
- **Ticks:** Apply acaricides or use integrated pest management strategies, including rotating pastures, to manage tick infestations.
- **Fleas and Lice:** Use livestock-specific insecticides or dips for effective control and maintain clean and dry living conditions for animals.
- **Mites (e.g., Sarcoptes and Demodex):** Treat affected animals with acaricides and isolate or cull severely infested individuals to prevent the spread.

- **Flies (e.g., Stable Flies and Horn Flies):** Use fly control methods such as fly traps, insecticide ear tags, or larvicides in manure management.
- **Liver Flukes:** Prevent liver fluke infection by managing water sources, controlling snail populations, and deworming livestock as necessary.
- **Lungworms:** Administer anthelmintic drugs targeting lungworms and practice good pasture management to reduce exposure.
- **Maggots (Myiasis):** Maintain good hygiene, clean wounds, and use fly repellents to prevent myiasis infestations.
- **Biting Midges (Culicoides):** Protect livestock from biting midges by using insect repellents and implementing measures to minimize breeding sites, such as standing water.
- **Babesia and Anaplasma (Tick-Borne Pathogens):** Control tick populations to reduce the risk of these bloodborne diseases and consider vaccination for some cases.
- **Scabies (Sarcoptic Mange):** Isolate and treat affected animals, as well as implement quarantine and treatment protocols to prevent the spread of scabies.
- **Foot Rot:** Isolate affected animals, maintain clean and dry environments, and use appropriate foot baths with disinfectants to manage foot rot.
- **Liver Abscesses:** Monitor and manage livestock nutrition, particularly in feedlot situations, to reduce the risk of liver abscesses.
- **Sheep Keds (Louse Flies):** Use insecticidal sprays, dips, or systemic insecticides to control sheep keds and ensure proper animal handling to minimize stress.
- **Cryptosporidium:** Improve sanitation, isolate infected animals, and provide supportive care to minimize the effects of cryptosporidiosis.
- **White Muscle Disease (Selenium and Vitamin E Deficiency):** Supplement livestock diets with selenium and vitamin E or provide free-choice mineral mixes to prevent this nutritional deficiency.
- **Bovine Tuberculosis:** Implement surveillance and testing programs, isolate and cull infected animals, and practice

biosecurity measures to control bovine tuberculosis.
- **Tapeworms**: Deworm livestock with products effective against tapeworms and use good pasture management to reduce tapeworm exposure.
- **Tsetse Flies (African Trypanosomiasis)**: Use insecticide-treated targets or traps to control tsetse fly populations and implement tsetse fly control programs in affected regions.
- **Ectoparasitic Mites in Chickens (e.g., Northern Fowl Mites)**: Isolate and treat affected birds, clean and disinfect housing, and use acaricides to control mite infestations.
- **Lambing/Kidding Sickness**: Provide appropriate nutrition to pregnant ewes and does, monitor for pregnancy toxemia, and provide supportive care during lambing or kidding.

Conclusion

One of the most critical takeaways of this book is the mantra of "Rotation, rotation, rotation." Implementing a regular rotation schedule for your livestock, moving them into new pastures and paddocks at least every 60 days, is the cornerstone of a successful parasite management strategy. Consider shortening the rotation period to every 30 days for even better results, and during outbreaks or high-load times, increasing the frequency becomes the best practice. The power of effective pasture management through rotation cannot be overstated, as it minimizes exposure to parasites and encourages healthier living conditions for your animals.

Beyond rotation, maintaining clean and debris-free stable, coop, or cage areas is equally important. Regular cleaning, especially after deworming times, is essential to prevent reinfection. Utilizing drying agents like lime and diatomaceous earth can further improve your efforts in creating an environment less favorable for parasites.

An often overlooked aspect of parasite control is the health of pastures and water systems. Keeping water sources clean and ensuring your livestock refrain from drinking from ponds, creeks, puddles, or other potentially contaminated water sources is key. During extremely wet conditions, exercise caution and avoid grazing livestock on pastures with widespread manure and short forage.

Another vital strategy is putting a structured approach to deworming in place. Administer antiparasitics for at least three consecutive days every month with housed animals. Follow this with a free-choice lick containing fresh ingredients or herbs, or run your livestock through a beneficial plant paddock for three additional days. By creating bulk mixes and boluses and procuring the most ingredients online, you can streamline deworming practices, making them more efficient and cost-effective.

Recognize that some animals are more susceptible to parasitic infestations than others. They may show signs of reduced vitality or sluggishness compared to other animals within the group. These individuals should be monitored closely, with regular fecal samples collected and analyzed. Consider culling these animals if feasible, especially if their susceptibility is due to poor eating habits or a lack of consumption of antiparasitic forages. Preventing the passage of these poor habits to their offspring is crucial for improving your herd or flock overall.

With these strategies in place, the need for frequent fecal tests will diminish, and the battle against internal parasites will no longer be a time-consuming chore. Your livestock will thrive in a healthier, more sustainable environment, and you'll have the peace of mind that comes with knowing that you've mastered the power of natural parasite management.

By embracing the practices outlined in this book and tailoring them to your specific circumstances and animals, you'll unlock the potential for a healthy, resilient herd or flock. With ongoing dedication to these strategies, you'll protect your livestock and build a future where the relentless pursuit of parasite management transforms into a harmonious and part-time aspect of your livestock care routine.

The well-being of your animals and the success of your livestock operation are within your reach. The journey may have started with questions and concerns, but now it ends with knowledge, empowerment, and a brighter future for your livestock and land.

Here's another book by Dion Rosser that you might like

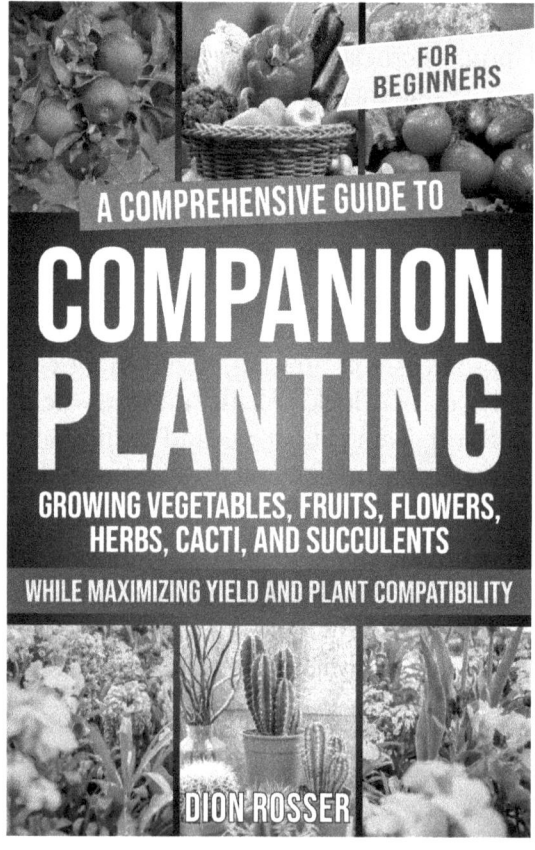

References

5 Tips to Raising Livestock for Food. (2015, April 1). Melissa K. Norris. https://melissaknorris.com/tips-to-raising-livestock/

abundantlifehomestead. (2018, June 3). Homemade Baby Food. Abundant Life Homestead. https://abundantlifehomestead.com/homemade-baby-food/

Affordable Ways to Build a House. (n.d.). Www.homesteadfunding.com. https://www.homesteadfunding.com/blogs/affordable-ways-to-build-a-house

Allen, M. (2021, August). The 6 environmental and health benefits of growing your own food. Www.thegardencontinuum.com. https://www.thegardencontinuum.com/blog/the-6-environmental-and-health-benefits-of-growing-your-own-food

Andrews (MA), U. of S. (n.d.). What to Consider When Choosing a Homestead. Treehugger. https://www.treehugger.com/what-consider-choosing-homestead-5111926

Anna. (2019, August 23). Canned Zucchini Spread (Ikra) Recipe. Northern Homestead. https://northernhomestead.com/canned-zucchini-spread-ikra-recipe/

Australia, E. B. S. (n.d.-a). Beekeeping Gear For Beginners. Ecrotek Beekeeping Supplies Australia. https://www.ecrotek.com.au/blogs/articles/beginner-beekeeping-gear

Australia, E. B. S. (n.d.-b). Protecting Your Bees From Diseases And Pests. Ecrotek Beekeeping Supplies Australia. https://www.ecrotek.com.au/blogs/articles/protecting-your-bees-from-diseases-and-pests

Bradley, K. (n.d.). How to Reduce Your Homestead's Carbon Footprint - Grit. Www.grit.com. https://www.grit.com/farm-and-garden/how-to-reduce-your-homesteads-carbon-footprint-zb0z2001/

Burton, L. (2018, December 19). Food Preservation Methods & Guidance. The Hub | High Speed Training. https://www.highspeedtraining.co.uk/hub/food-preservation-methods/

C, C. (n.d.). Winter food storage guide in a root cellar or other cellaring methods - The Home Preserving Bible. http://www.homepreservingbible.com/1057-winter-food-storage-in-a-root-cellar-and-other-preservation-methods/#:~:text=A%20root%20cellar%20is%20an

Choosing the Best Farm Livestock Animals to Raise - Mother Earth News. (2021, April 20). Www.motherearthnews.com. https://www.motherearthnews.com/homesteading-and-livestock/best-farm-livestock-animals-to-raise-zmaz82ndzgoe/

Curing and Smoking Meats for Home Food Preservation. (n.d.). Bradley Smoker USA. https://www.bradleysmoker.com/blogs/articles-smoking-guide/curing-and-smoking-meats-for-home-food-preservation

Dairy Cattle Farming - Raising Cows For Milk. (n.d.). In.virbac.com. https://in.virbac.com/cattle/health-care/dairy-cattle-farming-raising-cows-for-milk

deeannecurtis. (2023, October 13). What to Look for When Buying Land for a Homestead: Key Considerations for Prospective Buyers. https://hummingbird-acres.com/what-to-look-for-when-buying-land-for-a-homestead-key-considerations-for-prospective-buyers/

Dehydrating Food: Is It Good for You? (n.d.). WebMD. https://www.webmd.com/diet/dehydrating-food-good-for-you

Ethical and Sustainable Beekeeping. (2022, October 19). Beeautify. https://beeautify.com.au/blogs/beeautify-blog/ethical-and-sustainable-beekeeping

Extension | Canning Process. (n.d.). Extension.wvu.edu. https://extension.wvu.edu/food-health/home-food-preservation/canning/canning-process

Food Preservation: Making Pickled Products. (2017, November 29). NDSU Agriculture and Extension. https://www.ndsu.edu/agriculture/extension/publications/food-preservation-making-pickled-products

Gardening, E. @ M. (2018, October 1). Small Homestead Layout Design Plans. Misfit Gardening. https://misfitgardening.com/small-homestead-layout-design-plans/

Guidelines for Harvesting Vegetables | Piedmont Master Gardeners. (n.d.). Piedmontmastergardeners.org. https://piedmontmastergardeners.org/article/guidelines-for-harvesting-vegetables/#:~:text=Harvest%20with%20the%20right%20tools

Holte, L. (2021, April 21). 7 Tips for Keeping Happy and Healthy Bees. Miller Manufacturing Company Blog. https://www.miller-mfg.com/blog/7-tips-for-keeping-happy-and-healthy-bees/

Household, I. W. (2022, December 12). The Pros and Cons of Off-Grid Living. Utopia. https://utopia.org/guide/the-pros-and-cons-of-off-grid-living/

How to Assemble A Bee Hive Box or Super - A Beginner Beekeeper's Guide. (n.d.). Beverly Bees. https://www.beverlybees.com/beginner-beekeepers-guide/assemble-bee-hive-box-super/

How to Preserve Meat, Eggs & Dairy. (2021, August 13). Melissa K. Norris. https://melissaknorris.com/podcast/how-to-preserve-meat-eggs-dairy/

How to Reduce, Reuse, and Recycle on the Homestead. (2022, October 7). Homesteading.com. https://homesteading.com/reduce-reuse-recycle-homestead/

How waste reduction can help you have a more sustainable homestead. (2019, April 5). Hello Homestead. https://hellohomestead.com/how-waste-reduction-can-help-you-have-a-more-sustainable-homestead/

https://realestate.usnews.com/real-estate/articles/how-to-choose-the-right-homestead-property. (n.d.).

https://www.cnbc.com/2019/07/02/spending-2-hours-in-nature-per-week-can-make-you-happier-and-healthier.html#:~:text=Those%20who%20spent%20at%20least,nature%20had%20no%20further%20benefits.

Jack, S. (2023, September 27). Homestead Animal Husbandry: 10 Tips for Raising Healthy Livestock. Survival Jack. https://survivaljack.com/2023/09/homestead-animal-husbandry-10-tips-for-raising-healthy-livestock/

Joe. (2020, August 6). Raising Sheep for Wool | Sheep Wool Info. RaisingSheep.net. https://www.raisingsheep.net/raising-sheep-for-wool

June 15, B. D. U., & 2020. (n.d.). How to Dehydrate Fruits and Vegetables for a Healthy Snack. EatingWell. https://www.eatingwell.com/article/290910/how-to-dehydrate-fruits-and-vegetables-for-a-healthy-snack/

Kim. (2018, August 4). How To Start A Vegetable Garden For Beginners. Homestead Acres. https://www.homestead-acres.com/how-to-start-a-vegetable-garden-for-beginners/

Laessig, I. (2019, May 16). How to Make Sprouted Bread. Sunday Supper Movement. https://sundaysuppermovement.com/how-to-make-sprouted-bread/

Magazines, C. (2012, March 27). Milking techniques: the best habits. Farmer's Weekly. https://www.farmersweekly.co.za/farming-basics/how-to-livestock/milking-techniques/

Pallotta, N. (2020, June 7). Vegan strawberry jam. The Plant Based School. https://theplantbasedschool.com/easy-strawberry-jam/

Penn State Extension. (2019, March 13). Let's Preserve: Basics of Home Canning. Penn State Extension. https://extension.psu.edu/lets-preserve-basics-of-home-canning

Pinterest, camp cooking gear to help you enjoy great food in the great outdoors! (2020, May 21). The Ultimate Guide to Dehydrating Food. Fresh off the Grid. https://www.freshoffthegrid.com/dehydrating-food/

RecycleNation. (n.d.). Making Your Own Dairy Products for Beginners - RecycleNation. https://recyclenation.com/2015/07/making-your-own-dairy-products-for-beginners/

Rhodes, J. (2022, September 10). What Farm Animals Eat in a Day. Abundant Permaculture. https://abundantpermaculture.com/what-farm-animals-eat-in-a-day/

Rucker, B. (2022, January 15). How to Build a Shelter for Livestock on Your Homestead | Homesteading Info. Earthineer. https://earthineer.com/animals/how-to-build-a-shelter-for-livestock-on-your-homestead/

Shearing Tips to be a pro. (n.d.). Www.lister-Global.com. https://www.lister-global.com/news/shearingpro/#:~:text=Use%20your%20left%20hand%20to

Spicy Plum and Apple Chutney [Vegan]. (n.d.). One Green Planet. https://www.onegreenplanet.org/vegan-recipe/spicy-plum-and-apple-chutney/

Tamara. (2023, October 6). How to Make Tomato Sauce from Cherry Tomatoes. The Reid Homestead. https://thereidhomestead.com/roasted-tomato-sauce-from-cherry-tomatoes/

The Best Livestock To Raise On A Homestead. (n.d.). Farmer Boy. https://farmerboyag.com/blog/the-best-livestock-to-raise-on-a-homestead/

The Difference Between Pickling and Fermenting. (n.d.). Spicesinc.com. https://spicesinc.com/blogs/difference-between-pickling-and-fermenting#:~:text=An%20easy%20way%20to%20remember

Thomas, C. (2020, August 1). What To Do With Raw Milk + The Anatomy of Raw Milk. Homesteading Family. https://homesteadingfamily.com/what-to-do-with-raw-milk/

Types of Alternative Energy on a Homestead. (n.d.). Survivalist 101. https://survivalist101.com/tutorials/survivalist-homesteading-101/alternative-energy-sources/

University, U. S. (n.d.). Beekeeping | Extension. Extension.usu.edu. https://extension.usu.edu/beekeeping/learn/beginning-beekeeping/index

(N.d.-a). Nih.gov. https://www.ncbi.nlm.nih.gov/pmc/articles/PMC7767362/

(N.d.-b). Nih.gov. https://www.ncbi.nlm.nih.gov/pmc/articles/PMC5756309/#:~:text=Background%3A,activities%20including%20anti%2Dparasitic%20effect.

A brief history of parasitology. (2023, March 3). Veterinary Practice. https://www.veterinary-practice.com/article/history-of-parasitology

A guide to herbal remedies. (n.d.). Medlineplus.gov. https://medlineplus.gov/ency/patientinstructions/000868.htm

Ahmed, M., M.D. Laing, and I.V. Nsahlai. 2013. Studies on the ability of two isolates of Bacillus thuringiensis, an isolate of Clonostachys rosea f. rosea, and a diatomaceous earth product, to control gastrointestinal nematodes of sheep. Biocontrol Science and Technology.

Alok, A. (2015). Curcumin - pharmacological actions and its role in oral submucous fibrosis: A review. Journal of Clinical and Diagnostic Research: JCDR, 9(10), ZE01. https://doi.org/10.7860/jcdr/2015/13857.6552

Alzohairy, M. A. (2016). Therapeutics role ofAzadirachta indica(neem) and their active constituents in disease prevention and treatment. Evidence-Based Complementary and Alternative Medicine: eCAM, 2016, 1–11. https://doi.org/10.1155/2016/7382506

Amalraj, A., Pius, A., Gopi, S., & Gopi, S. (2017). Biological activities of curcuminoids, other biomolecules from turmeric and their derivatives - A review. Journal of Traditional and Complementary Medicine, 7(2), 205–233. https://doi.org/10.1016/j.jtcme.2016.05.005

Animal husbandry - Nature Neem. (n.d.). Natureneem.com. https://natureneem.com/en/solutions/animal-husbandry

Anthelmintic. (n.d.). Herbal Reality. https://www.herbalreality.com/western-action/anthelmintic/

Athanasiadou, S., Githiori, J., & Kyriazakis, I. (2007). Medicinal plants for helminth parasite control: facts and fiction. Animal: An International Journal of Animal Bioscience, 1(9), 1392–1400. https://doi.org/10.1017/s1751731107000730

Aylott, R. I. (2003). GIN | The Product and its Manufacture. In Encyclopedia of Food Sciences and Nutrition (pp. 2889–2893). Elsevier.

Azarpajouh, S. (2022, November 15). Breeding for parasite resistance in dairy cows. Dairy Global. https://www.dairyglobal.net/health-and-nutrition/health/breeding-for-parasite-resistance-in-dairy-cows/

Bang KS, Familton AS, Sykes AR (1990) Effect of copper oxide wire particle treatment on establishment of major gastrointestinal nematodes in lambs. Research in Veterinary Science 49:132-139

Barkley, M. (n.d.). Prevent parasites through grazing management. Psu.edu. https://extension.psu.edu/prevent-parasites-through-grazing-management

Beltran, M. A. G., & Martin, R. J. (2015, September 1). Home page. Doi.Org; unknown. https://doi.org/

Best management practices for pasture parasite management. (2019, May 6). Cornell University College of Veterinary Medicine. https://www.vet.cornell.edu/animal-health-diagnostic-center/programs/nyschap/modules-documents/best-management-practices-pasture-parasite-management

Better for animals. (n.d.). Soilassociation.org. https://www.soilassociation.org/take-action/organic-living/why-organic/better-for-animals/

Biosecurity, L. (n.d.-a). Histopathology sampling guide for livestock. Gov.au. https://www.agric.wa.gov.au/livestock-biosecurity/histopathology-sampling-guide-livestock

Biosecurity, L. (n.d.-b). Livestock disease veterinary sampling guide. Gov.au. https://www.agric.wa.gov.au/livestock-biosecurity/livestock-disease-veterinary-sampling-guide

Bissa, S., & Bohra, A. (2011). Antibacterial potential of pot marigold. Academicjournals.org. https://academicjournals.org/journal/JMA/article-full-text-pdf/F3AA1F49795

Bom Harris, D. V. M. (2020, September 1). Prevent parasites with pasture management. Old-Dominion-Vets. https://www.olddominionvets.com/post/prevent-parasites-with-pasture-management

Bosco, A., Prigioniero, A., Falzarano, A., Maurelli, M. P., Rinaldi, L., Cringoli, G., Quaranta, G., Claps, S., Sciarrillo, R., Guarino, C., & Scarano, P. (2023). Use of perennial plants in the fight against gastrointestinal nematodes of sheep. Frontiers in Parasitology, 2, 1186149. https://doi.org/10.3389/fpara.2023.1186149

Brief history herbal Medicine. (2018, May 16). Herbal Clinic - Swansea. https://www.herbalclinic-swansea.co.uk/herbal-medicine/a-brief-history-of-herbal-medicine/

Burke JM, Miller JE, Olcott DD, Olcott BM, Terrill TH (2004) Effect of copper oxide wire particles dosage and feed supplement level on Haemonchus contortus infection in lambs Veterinary Parasitology 123:235– 243

Cattle tick fever. (n.d.). MLA Corporate. https://www.mla.com.au/research-and-development/animal-health-welfare-and-biosecurity/parasites/identification/cattle-tick-fever/

Climate change and the expansion of animal and zoonotic diseases – what is the agency's contribution? (2021, June 2). Iaea.org. https://www.iaea.org/resources/news-article/climate-change-and-the-expansion-of-animal-and-zoonotic-diseases-what-is-the-agencys-contribution

Climate impacts on agriculture and food supply. (n.d.). Chicago.gov. https://climatechange.chicago.gov/climate-impacts/climate-impacts-agriculture-and-food-supply

Coates, J. (2012, October 25). The history and use of herbal medicine and its use today for pets. Petmd.com; PetMD. https://www.petmd.com/blogs/fullyvetted/2012/oct/history_and_use_of_herbal_medicine_and_use_in_pets-29279

coccidiosis. (n.d.). MLA Corporate. https://www.mla.com.au/research-and-development/animal-health-welfare-and-biosecurity/parasites/identification/coccidiosis/

Common cattle parasites. (2023, September 18). Texas A&M AgriLife Extension Service. https://agrilifeextension.tamu.edu/asset-external/common-cattle-parasites/

Common internal parasites of cattle. (n.d.). Missouri.edu. https://extension.missouri.edu/publications/g2130

D., l. L. H. (n.d.). Economic impact of gastrointestinal parasitism in Amazon buffalo far- Brazil. Embrapa.Br. https://www.alice.cnptia.embrapa.br/alice/bitstream/doc/403427/1/Economicimpact.pdf

Dai, Y.-L., Li, Y., Wang, Q., Niu, F.-J., Li, K.-W., Wang, Y.-Y., Wang, J., Zhou, C.-Z., & Gao, L.-N. (2022). Chamomile: A review of its traditional uses, chemical constituents, pharmacological activities and quality control studies. Molecules (Basel, Switzerland), 28(1), 133. https://doi.org/10.3390/molecules28010133

DIATOMACEOUS EARTH as an alternative treatment for internal parasites. (n.d.). SA Mohair Growers. https://www.angoras.co.za/article/diatomaceous-earth-as-an-alternative-treatment-for-internal-parasites

Diatomaceous Earth. (n.d.). Mdsmallruminant. https://www.sheepandgoat.com/de

Dotto, J. M., & Chacha, J. S. (2020). The potential of pumpkin seeds as a functional food ingredient: A review. Scientific African, 10(e00575), e00575. https://doi.org/10.1016/j.sciaf.2020.e00575

Eminov: Effect of certain pasture plants on gastrointesti... – Google Scholar. (n.d.). Google.Com. https://scholar.google.com/scholar_lookup?journal=Sov+Agric+Sci&title=Effect+of+certain+pasture+plants+on+gastrointestinal+nematodes+of+sheep&author=RS+Eminov&volume=1&publication_year=1982&pages=72-74&

ESCCAP. (n.d.). Glossary. Esccap.org. https://www.esccap.org/glossary/

Faculty By Department, & Find a Physician. (n.d.). Aloe. Rochester.edu. https://www.urmc.rochester.edu/encyclopedia/content.aspx?contenttypeid=19&contentid=Aloe

Farmacy. (2023, May 2). Managing your pasture to master the parasites. Farmacy. https://www.farmacy.co.uk/article/415-managing-your-pasture-to-master-the-parasites

Farmers Guardian. (2021, July 8). Management of pasture for parasite control. Farmersguardian.com. https://www.farmersguardian.com/sponsored/4092600/management-pasture-parasite-control

Ferguson, D., & Vogt, W. (2019, May 15). Fact sheet: Poisonous plants for cattle. Beefmagazine.com

flies. (n.d.). MLA Corporate. https://www.mla.com.au/research-and-development/animal-health-welfare-and-biosecurity/parasites/identification/flies/

Gastrointestinal worms. (n.d.). MLA Corporate. https://www.mla.com.au/research-and-development/animal-health-welfare-and-biosecurity/parasites/identification/gastrointestinal-worms/

Getting rid of intestinal parasites with diatomaceous earth. (n.d.). Www.Sassyorganics.Com.Au. https://www.sassyorganics.com.au https://www.sassyorganics.com.au/blog/our-blog/getting-rid-of-intestinal-parasites-with-diatomace/

Glossary. (n.d.-a). Cornell.edu. https://biocontrol.entomology.cornell.edu/glossary.php

Glossary. (n.d.-b). Ucanr.edu. https://ipm.ucanr.edu/PMG/glossary.html

Gupta. (2010). Chamomile: A herbal medicine of the past with a bright future (Review). Molecular Medicine Reports, 3(6), 895. https://doi.org/10.3892/mmr.2010.377

Hajaji, S., Alimi, D., Jabri, M. A., Abuseir, S., Gharbi, M., & Akkari, H. (2018). Anthelmintic activity of Tunisian chamomile (Matricaria recutita L.) against Haemonchus contortus. Journal of Helminthology, 92(2), 168–177. https://doi.org/10.1017/s0022149x17000396

Herbal medicine. (2021, September 24). Hopkinsmedicine.org. https://www.hopkinsmedicine.org/health/wellness-and-prevention/herbal-medicine

Herbal worming for cattle – McDowell's herbal treatments. (n.d.). McDowell's Herbal Treatments. https://www.mcdowellsherbal.com/success-stories-for-dogs/50-treatments/bovine-treatments/644-herbal-worming-for-cattle

Homemade herbal animal dewormer & tonic. (n.d.). Libertyhomesteadfarm.com. https://libertyhomesteadfarm.com/herbal-remedies/homemade-herbal-animal-dewormer-tonic/

How to work with your vet for the best farm outcomes. (2023, February 16). Pasture.Io. https://pasture.io/farm-animal-health/working-with-your-vet

Islam, M. S., & Rahman, M. M. (2016). Diatomaceous earth-induced alterations in the reproductive attributes in the housefly Musca domestica L. (Diptera: Muscidae). SSRN Electronic Journal, 96, 41241–41244. https://doi.org/10.2139/ssrn.3856328

Jaja, I., & Ungeviwa, P. (2022). A 6-year retrospective report of livestock parasitic diseases in the Eastern Cape Province, South Africa. Open Veterinary Journal, 12(2), 204. https://doi.org/10.5455/ovj.2022.v12.i2.8

Klasing, K. C., & Leshchinsky, T. V. (2000). Interactions between nutrition and immunity: Lessons from animal agriculture. In Nutrition and Immunology (pp. 363–373). Humana Press.

Kumar, N., Rao, T. K. S., Varghese, A., & Rathor, V. S. (2013). Internal parasite management in grazing livestock. Journal of Parasitic Diseases: Official Organ of the Indian Society for Parasitology, 37(2), 151–157. https://doi.org/10.1007/s12639-012-0215-z

Lefrançois, T., & Pineau, T. (2014). Public health and livestock: Emerging diseases in food animals. Animal Frontiers, 4(1), 4–6. https://doi.org/10.2527/af.2014-0001

Lice. (n.d.). MLA Corporate. https://www.mla.com.au/research-and-development/animal-health-welfare-and-biosecurity/parasites/identification/lice/

Liver fluke. (n.d.). MLA Corporate. https://www.mla.com.au/research-and-development/animal-health-welfare-and-biosecurity/parasites/identification/liver-fluke/

Livestock and poultry infectious diseases: Pathogenesis and immune mechanisms. (n.d.). Frontiersin.org. https://www.frontiersin.org/research-topics/47450/livestock-and-poultry-infectious-diseases-pathogenesis-and-immune-mechanisms

Livestock disease: Cause and control – Oklahoma state university. (2017, March 1). Okstate.edu. https://extension.okstate.edu/fact-sheets/livestock-disease-cause-and-control.html

Livestock management. (2018, August 28). Rodale Institute. https://rodaleinstitute.org/why-organic/organic-farming-practices/livestock-management/

Livestock parasites. (n.d.). Gov.au. https://www.agric.wa.gov.au/livestock-animals/livestock-management/livestock-parasites

M. WIEWIÓRA, M. ŁUKASIEWICZ, J. BARTOSIK, M. MAKARSKI, T. NIEMIEC. (2015). Diatomaceous earth in the prevention of worm infestation in purebred pigeons. Ann. Warsaw Univ. of Life Sci. – SGGW Animal Science, 54(2).

Mahleyuddin, N. N., Moshawih, S., Ming, L. C., Zulkifly, H. H., Kifli, N., Loy, M. J., Sarker, M. M. R., Al-Worafi, Y. M., Goh, B. H., Thuraisingam, S., & Goh, H. P. (2021). Coriandrum sativum L.: A review on ethnopharmacology, phytochemistry, and cardiovascular benefits. Molecules (Basel, Switzerland), 27(1), 209. https://doi.org/10.3390/molecules27010209

Mandal, S., & Mandal, M. (2015). Coriander (Coriandrum sativum L.) essential oil: Chemistry and biological activity. Asian Pacific Journal of Tropical Biomedicine, 5(6), 421–428. https://doi.org/10.1016/j.apjtb.2015.04.001

Mao, Q.-Q., Xu, X.-Y., Cao, S.-Y., Gan, R.-Y., Corke, H., Trust Beta, & Li, H.-B. (2019). Bioactive Compounds and Bioactivities of Ginger (Zingiber officinale Roscoe). Foods (Basel, Switzerland), 8(6), 185. https://doi.org/10.3390/foods8060185

Marcogliese, D. J. (2001). Implications of climate change for parasitism of animals in the aquatic environment. Canadian Journal of Zoology, 79(8), 1331–1352. https://doi.org/10.1139/z01-067

Marigold (Calendula). (2019, April 16). WholisticMatters. https://wholisticmatters.com/herb-detail/marigold-calendula/

Marosi, G., Szolnoki, B., Bocz, K., & Toldy, A. (2017). Fire-retardant recyclable and biobased polymer composites. In Novel Fire Retardant Polymers and Composite Materials (pp. 117–146). Elsevier.

Medicinal herb recipes for livestock. (n.d.). Sarahflackconsulting.com. https://www.sarahflackconsulting.com/articles/medicinal-herb-recipes-for-livestock/

Ndao, M. (2009). Diagnosis of parasitic diseases: Old and new approaches. Interdisciplinary Perspectives on Infectious Diseases, 2009, 1–15. https://doi.org/10.1155/2009/278246

Ntare, K. (n.d.). Natural herbs for treating Livestock. – Jaguza Farm Support. Jaguzafarm.com.

Nutrient management on livestock farms: Tips for feeding. (n.d.). Rutgers.edu. https://njaes.rutgers.edu/fs1064/

Parasites and Strategic Deworming. (n.d.). The College of Veterinary Medicine at Michigan State University. https://cvm.msu.edu/hospital/services/equine-services/for-owners/general-conditions-and-seeing-your-vet/parasites-and-strategic-deworming

Parasites, Diseases, and Control Measures. (n.d.). Usda.Gov. https://www.nal.usda.gov/exhibits/speccoll/exhibits/show/parasitic-diseases-with-econom/parasitic-diseases-with-econom

Parasites, diseases, and control measures. (n.d.). Usda.gov. https://www.nal.usda.gov/exhibits/speccoll/exhibits/show/parasitic-diseases-with-econom/parasitic-diseases-with-econom

Parasites, diseases, and control measures. (n.d.). Usda.gov. https://www.nal.usda.gov/exhibits/speccoll/exhibits/show/parasitic-diseases-with-econom/parasitic-diseases-with-econom

Parasites. (n.d.). MLA Corporate. https://www.mla.com.au/research-and-development/animal-health-welfare-and-biosecurity/parasites/

parasites. (n.d.). MLA Corporate. https://www.mla.com.au/research-and-development/animal-health-welfare-and-biosecurity/parasites/

Parasites. (n.d.). MLA Corporate. https://www.mla.com.au/research-and-development/animal-health-welfare-and-biosecurity/parasites/

Pasture management to control cattle worms. (2022, July 26). WormBoss. https://wormboss.com.au/management/non-chemical-worm-control-methods/pasture-management/

Pilarczyk, B., Tomza-Marciniak, A., Pilarczyk, R., Sadowska, N., Udała, J., & Kuba, J. (2022). The effect of season and meteorological conditions on parasite infection in farm-maintained mouflons (Ovis aries musimon). Journal of Parasitology Research, 2022, 1165782. https://doi.org/10.1155/2022/1165782

Plants poisonous to livestock. (n.d.). Missouri.edu. https://extension.missouri.edu/publications/g4970

Płoneczka-Janeczko, K., Szalińska, W., Otop, I., Piekarska, J., & Rypuła, K. (2023). Weather parameters as a predictive tool potentially allowing for better monitoring of dairy cattle against gastrointestinal parasites hazard. Scientific Reports, 13(1), 1–12. https://doi.org/10.1038/s41598-023-32890-0

Rahmani, A. H., Al shabrmi, F. M., & Aly, S. M. (2014). Active ingredients of ginger as potential candidates in the prevention and treatment of diseases via modulation of biological activities. International Journal of Physiology, Pathophysiology and Pharmacology, 6(2), 125.

Rahmann, G., & Seip, H. (n.d.). Bioactive forage and phytotherapy to cure and control endo-parasite diseases in sheep and goat farming systems – a review of current scientific knowledge. Orgprints.Org. https://orgprints.org/id/eprint/12976/1/181_Endoparasiten_Artikel_no_2_von_Rahmann_und_Seip.pdf

Richards, L. (2022, May 31). Wormwood: Uses, benefits, and risks. Medicalnewstoday.com. https://www.medicalnewstoday.com/articles/wormwood

Rizwan, H., Sajid, M., Shamim, A., Abbas, H., Qudoos, A., Maqbool, M., Malik, M., & Amin, Z. (2021). Sheep parasitism and its control by medicinal plants: A review. Parasitologists United Journal, 14(2), 112–121. https://doi.org/10.21608/puj.2021.70534.1114

Salehi, A., Razavi, M., & Vahedi Nouri, N. (2022). Seasonal prevalence of helminthic infections in the gastrointestinal tract of sheep in Mazandaran province, northern Iran. Journal of Parasitology Research, 2022, 7392801.

https://doi.org/10.1155/2022/7392801

Sandoval-Castro, C. A., Torres-Acosta, J. F. J., Hoste, H., Salem, A. Z. M., & Chan-Pérez, J. I. (2012). Using plant bioactive materials to control gastrointestinal tract helminths in livestock. Animal Feed Science and Technology, 176(1–4), 192–201. https://doi.org/10.1016/j.anifeedsci.2012.07.023

Sego, S. (2015, October 1). Managing benign prostatic hypertrophy with pumpkin seeds. Clinical Advisor. https://www.clinicaladvisor.com/home/features/alternative-meds-update/managing-benign-prostatic-hypertrophy-with-pumpkin-seeds/

Shang, A., Cao, S.-Y., Xu, X.-Y., Gan, R.-Y., Tang, G.-Y., Corke, H., Mavumengwana, V., & Li, H.-B. (2019). Bioactive compounds and biological functions of garlic (Allium sativum L.). Foods (Basel, Switzerland), 8(7), 246. https://doi.org/10.3390/foods8070246

Signs of worms in cattle. (2022, July 25). WormBoss. https://wormboss.com.au/about-worms/signs-of-worms/

Smith, J. (2019, June 20). Managing parasites in livestock. EcoFarming Daily. https://www.ecofarmingdaily.com/raise-healthy-livestock/cattle/managing-parasites-livestock/

Species and Life Cycles. (n.d.). Org.Uk. https://www.scops.org.uk/internal-parasites/worms/species-and-lifecycles/

Straub, C. (2023, March 12). Simple herbal remedies for your homestead animals. Biome Munch. https://biome-munch.com/2023/03/12/simple-herbal-remedies-for-your-homestead-animals/

Surjushe, A., Vasani, R., & Saple, D. G. (2008). Aloe vera: A short review. Indian Journal of Dermatology, 53(4), 163. https://doi.org/10.4103/0019-5154.44785

Sustainable parasite control for sheep and goats. (n.d.). Msstate.edu. http://extension.msstate.edu/publications/sustainable-parasite-control-for-sheep-and-goats

Sykes, A. R. (1994). Parasitism and production in farm animals. Animal Science (Penicuik, Scotland), 59(2), 155–172. https://doi.org/10.1017/s0003356100007649

Tavares, L., Santos, L., & Zapata Noreña, C. P. (2021). Bioactive compounds of garlic: A comprehensive review of encapsulation technologies, characterization of the encapsulated garlic compounds and their industrial applicability. Trends in Food Science & Technology, 114, 232–244. https://doi.org/10.1016/j.tifs.2021.05.019

The history of herbal medicine. (n.d.). New Chapter. https://www.newchapter.com/wellness-blog/the-history-of-herbal-medicine/

Theileriosis. (n.d.). MLA Corporate. https://www.mla.com.au/research-and-development/animal-health-welfare-and-biosecurity/parasites/identification/theileriosis/

Ticks. (n.d.). MLA Corporate. https://www.mla.com.au/research-and-development/animal-health-welfare-and-biosecurity/parasites/identification/ticks/

Tipsheet: Organic management of internal and external livestock parasites. (n.d.). Ncat.org. https://attra.ncat.org/publication/tipsheet-organic-management-of-internal-and-external-livestock-parasites/

Top natural remedy tips for common ailments in cattle. (n.d.). Farmcompare.Com. https://www.farmcompare.com/news/top-natural-remedy-tips-for-common-ailments-in-cattle

Toppo, A. (2018, December 10). Important livestock management tips for farmers. Krishi Jagran Media Group. https://krishijagran.com/animal-husbandry/important-livestock-management-tips-for-farmers/

Toxoplasma gondii – Learn About Parasites – Western College of Veterinary Medicine. (n.d.). Learn About Parasites. https://wcvm.usask.ca/learnaboutparasites/parasites/toxoplasma-gondii-zoonoses.php

Trichomoniasis. (n.d.). MLA Corporate. https://www.mla.com.au/research-and-development/animal-health-welfare-and-biosecurity/diseases/reproductive/trichomoniasis/

Vinje, E. (2014, February 17). The history of herbs and herbal medicine. Planet Natural. https://www.planetnatural.com/herb-gardening-guru/history/

Vlasova, A. N., & Saif, L. J. (2021). Bovine immunology: Implications for dairy cattle. Frontiers in Immunology, 12. https://doi.org/10.3389/fimmu.2021.643206

Vogt, W. (2023, April 27). Are parasite problems returning in cattle due to dewormer resistance? Beefmagazine.com. https://www.beefmagazine.com/cattle-disease/are-parasite-problems-returning-in-cattle-due-to-dewormer-resistance-

Waller, P. J., Bernes, G., Thamsborg, S. M., Sukura, A., Richter, S. H., Ingebrigtsen, K., & Höglund, J. (2001). Plants as DE-worming agents of livestock in the Nordic countries: Historical perspective, popular beliefs and prospects for the future. Acta Veterinaria Scandinavica, 42(1), 31. https://doi.org/10.1186/1751-0147-42-31

Website, N. H. S. (2022, October 19). Herbal medicines. Nhs.uk. https://www.nhs.uk/conditions/herbal-medicines/

Willis. (2020, April 8). Benefits of natural products for animals — Natural Animal Health. Natural Animal Health. https://www.naturalanimalhealth.co.uk/blog/benefits-natural-products-animals

Windon, R. G. (1990). Selective breeding for the control of nematodiasis in sheep: -EN- -FR- -ES-. Revue Scientifique et Technique (International Office of Epizootics), 9(2), 555–576. https://doi.org/10.20506/rst.9.2.496

Wong, C. (2004, April 26). Do parasite cleanses really work? Verywell Health. https://www.verywellhealth.com/natural-remedies-for-intestinal-parasites-88232

Wormers and the soil. (2020, December 18). FAS; Farm Advisory Scotland. https://www.fas.scot/article/wormers-and-the-soil/

www.ingramcontent.com/pod-product-compliance
Lightning Source LLC
Chambersburg PA
CBHW051855160426
43209CB00006B/1306